ALSO BY HAIM WATZMAN

Company C

A CRACK IN THE EARTH

A CRACK IN THE EARTH

A Journey up Israel's Rift Valley

HAIM WATZMAN

FARRAR, STRAUS AND GIROUX

NEW YORK

FARRAR, STRAUS AND GIROUX
19 Union Square West, New York 10003

Library of Congress Cataloging-in-Publication Data
Watzman, Haim.
    A crack in the earth / Haim Watzman.— 1st ed.
        p.   cm.
    Includes bibliographical references.
    ISBN-13: 978-0-374-13058-9 (hardcover : alk. paper)
    ISBN-10: 0-374-13058-2 (hardcover : alk. paper)
    1. Geology—Great Rift Valley—History.   2. Rifts (Geology)—Africa,
East.   3. Rifts (Geology)—Middle East   I. Title.

QE320.W38   2007
915.6904—dc22

                                                    2006021489

Designed by Gretchen Achilles

www.fsgbooks.com

1   3   5   7   9   10   8   6   4   2

Frontispiece: Shaded relief map based upon the 25m Digital Terrain Model
(DMT) of Israel, produced by Dr. John K. Hall, Geological Survey of Israel.
Reprinted with permission from Plate XI of the *Geological Framework of the
Levant*, Volumes I and II, edited by V. A. Krasheninnikov and J. K. Hall,
Historical Productions-Hall, Jerusalem, 2005.

FOR JUNE AND SANFORD WATZMAN,

MY PARENTS AND TEACHERS

For I have learned
To look on nature, not as in the hour
Of thoughtless youth; but hearing oftentimes
The still, sad music of humanity,
Nor harsh nor grating, though of ample power
To chasten and subdue. And I have felt
A presence that disturbs me with the joy
Of elevated thoughts; a sense sublime
Of something far more deeply interfused,
Whose dwelling is the light of setting suns,
And the round ocean and the living air,
And the blue sky, and in the mind of man . . .
—WILLIAM WORDSWORTH,
"Lines Composed a Few Miles Above Tintern Abbey
on Revisiting the Banks of the Wye
During a Tour, July 13, 1798"

The rules of behavior regarding release from vows float in the air
They have nothing to attach to;
The rules of behavior regarding the Sabbath, the feast of offerings,
     and inadvertent benefit from dedicated sacrifices
Are like mountains hanging by a hair
Since they have but little written in the Law and many rules;
Jurisprudence, the Temple service, purity and impurity, and
     forbidden unions
Have something to depend on—
They themselves are facts of the Law.
—MISHNAH HAGIGAH 1:8

CONTENTS

PART I

FACTS ON THE GROUND

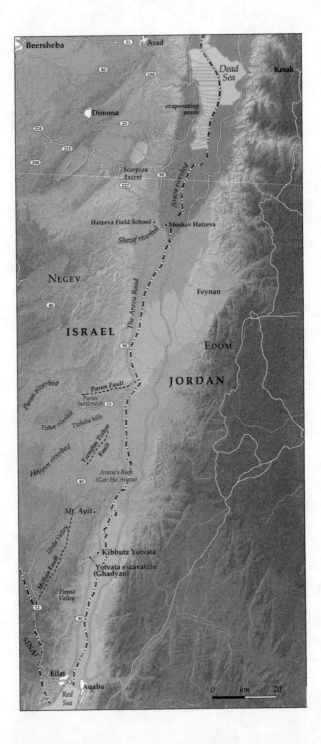

Beersheba
31 Arad
Dead
Sea
Kerak
90
258
evaporating
pools
Dimona
25
224
225
204
Scorpion
Ascent
90
227
Hatzeva Field School
Moshav Hatzeva
Shezaf riverbed
NEGEV
Feynan
40
ISRAEL
EDOM
90
Paran riverbed
Paran Fault
JORDAN
Paran
bottleneck 13
Tzihor riverbed
Tzehiha hills
Hayyun riverbed
Tzantfim Tzihor
Fault
Arava's Back
(Gav Ha'Arava)
40
Mt. Ayit
Uvda Valley
Kibbutz Yotvata
Yotvata excavation
(Ghadyan)
Milkan Fault
Tinna
Valley
12
90
SINAI
Eilat
Aqaba
0   km   20
Red
Sea

At the end of the promenade, past Herod's, over a rainbow bridge, the swanky stores of Eilat's hotel strip abruptly vanish. All along the beachfront, high-rise heavens with angled windows and palm-ringed swimming pools gaze down an arm of ocean that reaches out to them from the Arabian Sea. Only the plebeian strip of the promenade's fast-food stands and street vendors remind four- and five-star vacationers that the luxury they enjoy is tenuous and temporary. The gaudiest tower of them all is the faux palace that bears the name of the ancient East's greatest manipulator, madman, and master builder. Its rococo extravagance, all arches framed by columns and crowned by moon-bright domes, would have deeply offended the easily offended king's classical sensibilities. Herod's ("Where the Legend Comes Alive") is a temple of earthly delights that offers all-inclusive vacations of endless meals, celebrity shows, and classy boutiques. It would have enraged the great king so much that he'd probably have murdered yet another of his sons. It's not my kind of place. I'm staying at the youth hostel way down past the other end of the promenade, across the street from where the shoreline turns south toward Egypt and Africa.

But the epicurean paradise ends at Herod's eastern wall. Beyond it is a placid, silent canal spanned by an unlit convex bridge. On it, a few middle-aged fishermen cast their lines out, observed by their wives and by a silent, hungry cat. I step off the bridge into the planet's natural terrain. The elements rule the October night. Sand, sea, stone, and sky—the loneliness in which God resides. Thousands of stars suppressed by the brash promenade streetlamps reappear; the Dolphin and the Water Carrier hover to my right, over the sea; Polaris, low on my left, marks Route 90, the northward path of the two-week trip that I began four days ago. The beach is sandy and largely deserted, except for a couple of cars and a clapboard shack, which emits some light and the lilt of songs with Hebrew lyrics and Arab melodies.

I stand at the landfall of a great rift valley, a crack in the earth's crust that begins where the Indian Ocean's waters mix with those of the Gulf of Aden. It heads west by northwest, turns more sharply to the northwest, and at the Strait of Tiran, where the Sinai Peninsula comes to a point, it takes another turn and heads nearly due north before ending in the mountains of Anatolia. This rift is one of the globe's largest features, clearly visible from space, and I live on its edge. It forms an intricate landscape that makes the human soul turn end over end in wonder—even in people who are sure they have no organ by that name. One would have to be an automaton not to stand in awe of the God who designed it. Or so I felt when I first viewed the rift three decades ago.

In fact, we needn't call upon God to explain either the lay of the landscape or its origin. The rift is a geological fact, the product of enormous forces operating inside the globe, and it would exist even if there were no humans to observe it. Yet humans have been a part of it nearly since there were humans; the section I will travel, from the Red Sea north to the mountains of Syria, served as a corridor through which prehistoric humankind passed on its way out of Africa to colonize Asia and Europe. From that time on, they have left their mark on the valley, and it has marked their minds.

Now, even in satellite photographs, the rift cannot be seen pristinely. The light and heat emitted by Eilat and its Jordanian sister city, Aqaba, by Jerusalem, and by Tiberias on the shore of Lake Kinneret—the Sea of Galilee—stain the landscape as seen from outer space. Tiny Qaroun Lake in Lebanon's Bekaa Valley has a ruler-straight southern bank that nature could not have fashioned. It marks a dam and shows that the lake, which I rode past time and again when I served as an occupying soldier in Lebanon in the early 1980s, is man-made.

Here on earth, with my own eyes, I can see the valley only from its edge or within. Like a worker ant who climbs a blade of grass to get a better view of the hollow in which she will spend her brief life, I must use my mind as a ladder, in an attempt to grasp this great geological object. But even this is no simple matter, for humankind has overlaid the geology not just with cities, dams, fields, and roads but also with history and biography and meanings.

I have lived, traveled, and soldiered up and down the Israeli side of the rift valley in the twenty-seven years since this country became my home. For many of my fellow citizens, the greater part of the valley is a border that seems natural. But the location and nature of that border have been challenged by peace efforts and the winds of war alike. Israel's government is, at the time of my trip, seeking to redraw the country's boundaries in the Gaza Strip. Israelis who live in the Jordan Valley fear that a revision of the border along which they live is not long to follow. October of the year 2004 is thus an opportune time for me to travel my part of the rift and to see it as it is, but also as it signifies. Along my way I will meet geologists, biologists, and archaeologists who study the physical facts of the rift valley. I will speak to people who live and work in the valley and for whom it represents the fulfillment, or disappointment, of an ideal. And I will encounter others who see the rift through the fun-house mirror of myth, in which stories skip over the landscape and where human beings themselves are mysteries.

I walk across the sand and around the shack, where a brown picnic table, a green picnic table, and a table topped with yellow Formica stand on a concrete platform. Behind them is a countertop with a big sign next to it: "Fishermen's Snack Bar, Presenting: fishing line, sinkers, bait, and fishing lessons. Ice cream, hot and cold drinks."

Around the brown table, the one on the Red Sea side, sit five fishermen drinking beer and Coke. A woman with a weathered face and a soft smile sits with them; they glance at me but offer no greetings. The potent smell of raw fish pervades the patio, which is roofed with reeds and canes. A hundred yards past the snack bar is the frontier: a fence, a line of sandbags, a guard post, and a Jordanian flag on a high pole with a blinking red light on top. Beyond that are the lights of Aqaba, Eilat's sister city in the Hashemite Kingdom of Jordan.

Avraham, the dark-haired, bushy-eyed proprietor, eyes me. I'm a stranger intruding into his territory, and he doesn't believe that all I want is a glass of tea with *nana*, the Middle Eastern variety of mint. He looks beyond me to the yellow table, where I've already set up my PalmPilot and unfolded my keyboard, which admittedly look out of place. My second request seems to convince him, however. He makes me the tea and stirs in the spoonful of sugar I ask for. He clearly believes it's not enough. He asks if I'm here on vacation, and I say no, for work. I could come to the Moroccan synagogue tomorrow night for Sabbath services, he suggests, but I tell him that tomorrow I'm heading back to Jerusalem to spend the weekend with my family. I take the glass and saucer from him; the mint leaves circle slowly in the amber liquid. When I turn toward my table, he calls me back, places three gratis homemade cookies on the saucer, and offers some advice:

"Watch your pockets."

I sip my tea and gaze south, down the narrow finger of the Gulf of Eilat, called the Gulf of Aqaba if you live on the other side. The city across the way distracts my eye. It's practically a mirror image of Eilat, but an image enlarged by a magnifying mirror—it extends farther south along the shoreline, into the mountains, and north into the Arava plain. In fact, it's really Eilat that is the image in the mirror: historically, Aqaba is the site of human habitation at the north end of the gulf. The east side of the littoral gets more rain and has more sources of fresh water than modern Eilat, which was a tiny fishing village called Um Rash Rash before Israeli forces reached it during the War of Independence and made it the country's outlet to the Red Sea. Aqaba, known as Aylah to the Byzantine Christians and early Muslims, was the Levantine seaport for commerce with Arabia, Yemen, and India beyond.

Sometime during the period of Muslim rule, perhaps in the early 700s but more likely in the mid-800s, a man named Rabi, son of Qays, son of Yazid al-Ghassani, passed through the port town on his way north. A former highway robber, he was now superior of the Santa Katarina monastery at Mount Sinai. The Muslim rulers of Sinai had raised the monastery's taxes. Al-Ghassani, known to his acolytes as 'Abd al-Masih, the slave of the Messiah, was on his way to Ramla, the Muslim capital in the coastal plain below the Judean highlands, to plead for consideration. The Aramaic-speaking desert ascetics had probably made him their leader with this eventuality in mind, for 'Abd al-Masih was a native speaker of Arabic, the rulers' own language. The mission was an act of considerable altruism on al-Ghassani's part, for he was a former Muslim, and Muslim rulers punished apostasy with death. The young Santa Katarina monk who set the story down in broken Arabic a generation later tells us that 'Abd al-Masih had once before deliberately sought out martyrdom.

Amotz Zahavi, a biologist and evolutionary theorist whom I met the previous Sunday at the Hatzeva Field School, just off Route 90 a bit south of the Dead Sea, thinks he can explain why living organisms, man included, are sometimes willing to take risks, and even put their lives in danger, for the sake of others. From his observations of the social groups formed by birds who live in a riverbed near Hatzeva, he deduces that animals risk their lives to gain status and mating opportunities. With his status secure and his celibacy maintained by vow and power of will, 'Abd al-Masih doesn't seem to fit the theory. What, then, can explain his trip through Aylah?

In 1182, Renaud of Châtillon, a knight of the Second Crusade, former prince consort of Antioch, sacker of Cyprus, lord of the great

Crusader keeps in the Edom highlands that tower east of the Arava plain, built a seagoing fleet at the landlocked desert castle of Kerak. He had his men transport the disassembled boats 125 miles overland to the seashore, where he assembled and launched them, besieging the island of al-Qureiya, nine miles south; the Crusaders called this island Île de Graye. By controlling the roads at the northern tip of the sea, he could cut off the Muslim west, centered in Egypt, from the Muslim east in the Levant and Arabia. The Christians would then be able to charge tolls from Muslim caravans and pilgrims seeking to cross from one side of their world to the other, particularly pilgrims on their way to Mecca. Renaud's forces proceeded to plunder and pillage the Arabian coast. All this the prince consort had done without consulting and against the wishes of his nominal sovereign, King Baldwin. Baldwin was in the midst of consolidating and fortifying the Crusader state against the impending threat of the Muslim force then mustering in Syria under the command of a young and ambitious leader named Salah a-Din—the man whom the Franks called Saladin. Controlling the roads, however, was not enough for Renaud. Before the Muslim forces regrouped, he launched a raid across the gulf and deep into Arabia. By the time the Muslims were able to halt his advance, he was close to attacking Medina and Mecca, a knight of Jesus on violent pilgrimage to the shrines of Allah.

It's been two thousand years since the birth of Christianity. If each day in those two millennia had itself lasted a millennium, the church would then be as ancient as the rocks in the mountains on either side of me. They date to the Precambrian and are 570 million to 1 billion years old. The earth took form about 4.5 billion years ago, so a 1-billion-year-old rock was born after the earth had lived through seven-ninths of its history. A rock from the early Cambrian period, when fossils of animals larger than a single cell suddenly appear, was born when the earth was nearly eight-ninths of its present age.

Music blares from a boom box; it's Sarit Hadad, a popular singer in the genre of Hebrew pop songs in Arabic arrangements. Hadad sings blunt, prosaic lyrics about the sisterhood of the housing projects. It's her "Sea of Love": "Tell me, what'd you give me/What have I got from you?/No past and no future/I don't need you." It's not the cleaned-up version you hear on the radio but a recording of a live performance, with loud, whining strings and massive parallel chords, offering no quarter to the Western ear. A little girl, maybe four years old, with dark, plump cheeks, dances on the green table as the woman sitting with the fisherman applauds. Avraham joins them. I overhear an occasional word: "disengagement," "Gaza," "Sharon," "asshole." A Jordanian oil tanker floats out at sea, continuing the straight line of the fence in the sand.

Geologists don't know for sure where it all began. The mountains, whose reddish cast is indiscernible in the dark, are part of an ancient, shattered craton, one of the primal cores around which the continents coalesced. Between 580 and 600 million years ago, as those first continents collided with one another to form the supercontinent of Pangaea, the pressure caused high mountains to rise.

The next 300 million years, the Paleozoic era, are represented by Nubian sandstone. This rust-colored rock formed during a long period when the land I now stand on was a desert on the edge of an ocean, part of an uninterrupted Arabo-Nubian landmass that had no Red Sea in the middle. During the nearly 200 million years that followed, the Mesozoic era, the age of the dinosaurs, the Levant was mostly under water. Pangaea split up, with a sea called the Tethys forming between its fragments and flooding the former desert. The sea's dominion continued as India, once attached to Africa, swung north and east to collide with Asia 50 million years ago, pushing up the Himalayas. Antarctica disconnected from South America and moved toward the South Pole. The sea current that formed around the southernmost continent cooled the planet, making sea levels drop, and the Levant emerged from below the waves. The Middle East and Arabia were, at this point, still part of Africa.

Then, about 25 million years ago, in what geologists call the Miocene epoch—the time when grasslands, camels, and anthropoid apes first appeared—Africa cracked and Arabia began to move north-ward. An upwelling from the mantle, the huge, thick layer of viscous rock that lies below the earth's crust, stretched and broke it and caused it to collapse. The collapse formed the basin that was the first incarnation of the Red Sea. Volcanoes spewed lava on a line running from southeast to northwest, straight up to what is now the Gulf of Suez. The seafloor spread, creating new crust out of magma. Twenty million years ago the direction of the pressures changed. They took a right turn at Sinai's southern apex, leaving the Gulf of Suez in embryo form. The crack began to run northward. The land to the east of the fault line—just a line then, not a valley, crept north rela-tive to the west. Later, about 5 million years ago, the two sides be-gan to spread apart, forming the gulf before me and the rift valley that stretches to the north behind my back.

A car inches over the sand, its headlights reflecting in the quiet waters. The Red Sea is narrow, a passageway rather than an ex-panse, so it does not have big waves. It does not pound its shore. It laps it, like a kitten.

I ponder the border post on my left. My writer's instinct tells me to walk over to it and experience it close up, in whatever way one experiences a border post. But there are a few dark figures by the car and elsewhere on the sand—perhaps fishermen, perhaps beach-combers, but perhaps the pickpockets Avraham alluded to. It's Thurs-day night, and I am anxious to get home tomorrow. Israel has already reverted to standard time, and the day is short. I have a man to talk to here in the morning and a long drive home—a determined motorist can do it in four hours, but I am an occasional driver who can't stay on the road for more than an hour and a half without tak-ing a break. If I don't get home by the beginning of the Sabbath, just before sunset, I don't get home at all, since the laws of Orthodox Ju-daism don't allow driving once the Sabbath has begun.

I'm close to the end of the first week of my two-week trip through the Dead Sea Rift, or Dead Sea Transform, or Dead Sea De-pression. The exact name depends on which geological faction you belong to. The vast majority of geologists maintain that the evidence decisively supports the theory that the long, narrow valley to my north is a crack in the earth's crust, a fault that separates two tectonic plates, two pieces of the planet's surface. So I'll call it a rift.

The Dead Sea Rift is a narrow lowland with walls of mountains on either side. Geologically, it is a single entity. Here where I sit, and northward to just south of Lake Kinneret, the Sea of Galilee, it is also a border between two countries, Israel and Jordan. From here to the north end of the Dead Sea, and then again from just south of the town of Beit She'an to just south of Lake Kinneret, it is also an ethnic, religious, and cultural border, separating Hebrew speakers from Arabic speakers (sharply), Jews from Muslims (not quite as sharply), and Western ways from Oriental ways (even less sharply). Between the north of the Dead Sea and Beit She'an, where the rift intersects the liminal West Bank, the division is hazy. There are no Jews or Israelis on the east side of the rift, but there are many Palestinian Arabs, mostly Muslim, on the west side.

In theory, the border before me is permeable. For the last thirty-five years, it has been a largely quiet frontier. For the last decade, the two countries have officially been at peace. Israelis can go to Jordan, and Jordanians can come to Israel, but even during the few good years after the Israel-Jordan peace treaty of 1994, the movement was modest in comparison with, say, the hundreds of thousands who cross the U.S.-Mexican border daily. Here, the traffic was mostly tourists. Some Israelis hopped over to Jordan for short guided tours, but very few Jordanians crossed to travel in Israel.

Since the outbreak of the second Palestinian uprising, the al-Aqsa Intifada, in 2000, travel has been limited for the most part to a handful of professionals. The archaeologists and geologists I've met this week go to Jordan occasionally for conferences or to conduct research. At the state-of-the-art crossing point just north of the twin ports, which I visited some months later with an international group of scholars who study borders, an Israeli or a Jordanian can receive a visa on the spot and drive freely into the neighboring country.

Hardly anyone does so; the terminal's workers and guards look horribly bored. An Estonian professor and his wife decide to forgo the remainder of the organized tour and spend the rest of the day in Aqaba. With the exception of a few Jordanian laborers who cross over daily to work on construction projects in Eilat, they are the only people to cross that day.

A month before my October trip, the Israeli government issued a travel advisory against going to Jordan and Egypt. Two weeks later, the warning turned out to be something of substance: a suicide car bomber blew up a hotel in Taba, just over the border in Egypt, where Israeli families were spending the Sukkot holiday. On top of that, it's now the Muslim holy month of Ramadan, a good time for crimes of religious passion. I am not going to Jordan.

In the north, the rift valley continues into Lebanon and Syria, where I can't go even if I want to. These two countries are still officially at war with Israel. A hot phase of that war in the early 1980s gave me an opportunity to see the Lebanese extension of the rift, the Bekaa Valley, as an Israeli soldier, but I won't be going there on this trip. The border is not a feature of the earth's crust, but it is a greater barrier than the mountains and the seas I drive by and over during these two weeks.

I bid farewell to Avraham and the parliament of fishermen and cross back over the bridge into Eilat's flashy culture. It's off-season, so there are more hawkers than browsers. The fabulously dressed saleswomen in the fancy stores stand idly by their cash registers, looking like they'd rather be at home, wearing something more comfortable. The jewelry and art vendors along the promenade are in good spirits, trading jokes and trying to draw in the few shoppers who pass by. Apparently, a city bylaw requires that trinket vendors of both sexes be thin, long-haired, and possessed of at least three facial piercings and two tattoos. They are forbidden to wear sleeves. A different set of rules seems to govern the food business. Falafel stands are run by youths of sturdy frame, baguette and sandwich shops by middle-aged, overweight, and balding men. Cold drink concessions go to languorous, artificially blond, thirty-nine-year-old women, who are nightly inspected to ensure that they have applied too much makeup.

The walkers on the promenade are not buying much. Middle-aged Israeli couples form the plurality; they have worked through the holidays and saved vacation time so as to take advantage of off-season bargains at the fancy hotels. Two groups of retirement-age Russian immigrants pass by, led by tour guides. A scattering of high school guys on the make and girls looking for adventure have come down from the center of the country for a long weekend, skipping Friday classes. A dark-eyed woman with the long sleeves and head kerchief that identify her as ultra-Orthodox begs for coins. An emaciated man rides a three-wheeled cycle cart up and down the sidewalk, picking up deposit bottles; an Arab kid, eighteen or nineteen, laughs behind a wheelbarrow, pushing a friend into the service entrance of a Sheraton.

I stop at a candy store to buy M&M's for my kids. There are piles of unsold newspapers left over from the morning. Visitors to Eilat do not want to think about politics any more than they do history or geology. A fierce debate rages in the country north of here. Prime Minister Ariel Sharon wants to withdraw the army and all Israeli settlers from the Gaza Strip, leaving the Palestinians to fend for themselves behind the fence that surrounds that poverty-stricken, densely populated, and intensely Islamic territory. The Knesset, Israel's parliament, will vote on the policy next Tuesday. Earlier this week a group of Zionist rabbis called on religious soldiers to refuse to evacuate Israeli settlers. Yesterday another group of Zionist rabbis called on religious soldiers to obey orders, even if they object to the policy. These are precisely the kinds of things that people come to Eilat to escape.

Geologists tell different stories about what has been happening here for the last 5 million years. They examine the valley and the hilly plateaus on either side of it and compare them with similar formations around the world. Their aim is to reconstruct the process that created this landscape, and to predict where it will go from here.

From where I stand, I see a flat plain, slightly ascending to the north, framed by the dark outlines of the mountains. The line of the warmish salt water behind me is the current sea-level line, where the water intruding from the Indian Ocean stops. But the plain I see to the north continues south, under the water.

When I fly over the length of the Arava on my way home from my trip with the border scholars, I can see clearly that the plain is not continuous. In fact, it's a pockmarked series of basins, separated by ridges. Maps show three such basins on the floor of the Gulf of Aqaba. The northernmost of these does not end where the sea ends on the Eilat beach but actually stretches many miles to the north. It ends at a rise called the Arava's Back, the water divide between the Red and Dead seas. The Dead Sea and the lower Jordan valley to the north are another basin. The Beit She'an valley, the upper Jordan Valley, and Lake Kinneret compose another. The Hula Valley, in Israel's northern panhandle, is yet another, and the Ayyun valley in southern Lebanon continues the chain.

Many early geologists theorized that the valley is a graben. The word, like some of the earliest geologists, comes from Germany, and it means a piece of terrain that has collapsed between two or more faults and now lies low between higher ground. Still others said that something else was also going on—not just stretching but also slipping and sliding. In this view, the crack was created as the land on its east side moved north relative to the land on its west side.

One of the strongest pieces of evidence for this latter view is the fact that the rocks on either side of the valley don't match. If the valley is something that collapsed because of stretching, one would expect to find the same kinds of rocks in parallel positions on each side. Instead, rock formations on the west side seem to match rock formations that lie farther north on the east side. There are copper and manganese deposits in the mountains on the west side of the valley at Timna, about fifteen miles north of Eilat. If you draw a line due east, you find no matching deposits in the mountains on the other side. But there are copper and manganese deposits on the east side sixty-five miles to the north, east of the Dead Sea. The sixty-five-mile figure also emerges from other, independent pieces of evidence taken from the Red Sea floor and from the channels of ancient rivers that flowed from Jordan into the Mediterranean before the rift existed.

Grabens are caused by stretching, when two pieces of the earth's surface move apart. They may be the first stage in the birth of an ocean. When Pangaea pulled apart and the Atlantic Ocean began to open up between North America and Eurasia, the crust was stretched and magma welled up between them. It formed new seafloor, an upwelling that continues to this day in the Atlantic's mid-ocean mountain ridge. The same process is now going on in the Red Sea, which geologically is an embryonic ocean. Most everyone agrees that the two sides of the rift valley have been pulled apart. But if this were all that was happening between Israel and Jordan, the break ought to be more or less even. Instead, there are those basins and ridges.

Take a sheet of paper. Use a pair of scissors to cut a zigzag up the middle of the sheet. Hold the two halves of the cut paper together and then slowly begin to move the right piece upward. What do you see? A series of diamond-shaped holes, separated by places where the halves overlap. Basins and ridges.

This evidence has been enough to convince most geologists that the valley is a transform fault, like the San Andreas Fault in California. In a transform fault, sometimes called a strike-slip fault, two pieces of crust move alongside each other. In this case, the east side of the valley seems to be moving north relative to the west side. But some geologists hold out for explanations based on stretching. They have some points in their favor: the deposits that match over a sixty-five-mile jump north are evident in the south, but north of the Dead Sea the evidence is much more ambiguous. The fault line, clear in the south, is much more complex in the north, and in some places, not visible at all.

I've reached the end of the promenade, where Route 90 runs past the Eilat mall. My youth hostel room is the southernmost point in my trip. From here, I head north.

At the line on the slope where Eilat's high-rises end and bare mountain begins, Uzi Avner has erected shrines to gods whose names he does not know. In his former capacity as the Israel Antiquities Authority's point man in the Eilat region, Avner oversaw salvage excavations before construction began on Eilat's new neighborhoods. He found altars and standing stones and moved them up the hill, labeling them carefully.

Thirty-five miles up the road, around the air force base in the Ovdat Valley, he discovered the outlines of what he believes were places of worship thousands of years ago. Open-air enclosures are marked off by lines of stones. There is a hole in the ground and some large, jutting rocks at one end. Avner identifies the hole as a sunken altar and the enclosure as a temple. At one site, outside the enclosure, are lines of rocks that trace the shapes of animals—an oryx and two leopards.

My guide on the day I see them is Tali Gini of the Israel Antiquities Authority. She doesn't think much of the stone pictures. "There are two kinds of rocks in the drawings," she points out. The long, thin ones, she tells me, lie where Avner found them. He has filled out missing parts of the outlines with smaller pebbles. In the case of the oryx, the small pebbles are most of the picture, and it's hard to see how Avner knew that the rock artist intended to draw this particular animal, or any animal at all. She also points out that she's seen similar stone drawings at an early Islamic site, so she doesn't believe Avner can be sure that his connect-the-dot animals are in fact connected to the adjacent Neolithic structure. She accepts

Avner's dating of the structure, then, but not necessarily his hypothesis about its function.

It's an old saw that an ancient temple is a structure that archaeologists can't identify and a cult object is anything man-made that seems to have no practical use. Avner identifies a lot of temples and cult objects, and he takes flak about it from some of his colleagues. I suspect, after talking to some of those colleagues, that the flak is not only a matter of theoretical difference. Among tourists, students, and lay enthusiasts of archaeology, Avner has a well-deserved reputation for being a stimulating and compelling speaker, guide, and teacher. Among his colleagues in the field, he has no less a reputation for often being difficult to work with.

Gini thinks Avner has taken far too much liberty with the facts on the ground. He has filled out stone drawings, stood up rocks he found lying flat or leaning on other rocks, and re-created altars out of jumbles of stony debris. Some archaeologists are purists—they believe that ancient sites should be left as untouched as excavation permits. Others want to reconfigure the sites according to their interpretations. Avner maintains that he is cautious and conservative in his reconstructions. He rights only stones that have clearly fallen down. But he can't deny that erecting a stone is an act of interpretation.

I'm a journalist by training, and the skepticism that is the journalist's most important professional tool once led me to assume that any scholar who's at odds with his colleagues must be onto something. If the academic establishment says it's trash, it must be a revolutionary truth, a threat to fogies who wish to protect their turf from new ideas. Over the years, though, I've learned to direct my skepticism at the lone wolves as well. New ideas sometimes need people with daring to champion them, but the daring itself is no guarantee of truth.

Uzi Avner has patiently explained his views to me, mustered his evidence, and made his case. I don't have the professional expertise to judge competing archaeological claims. My field is telling stories, and he has an interesting story to tell.

Avner has graying hair and a high forehead. He receives me in the study of his cozy row house in a neighborhood up the slope. It's an early-morning meeting, so his family is still asleep. I see from his computer screen that, before I arrived, he was working on a grant application.

"The Bible talks a lot about pillars—about standing stones," Avner tells me. "In the Five Books of Moses there are seventeen verses that talk about erecting standing stones or breaking the stones of other people. All in all, the Bible doesn't like them. If the Israelites were constantly being told to knock down standing stones, it means there must have been lots of stones to knock down. That reality as reflected in the biblical text is confirmed by archaeology. The desert is full of standing stones. I've documented more than 230 sites in the Negev desert with standing stones. You find them all over the world. What are they?"

He goes on to tell me that nearly all ancient peoples erected large stones. Ancient texts and the practices of tribal societies today give us clues to their use. Some, such as the stelae of ancient Egypt and Mesopotamia, were erected in order to commemorate events or people. Such stones generally carry inscriptions or depictions of the events or people in whose honor they were erected. Others serve as witnesses of treaties or vows; we see this in the Bible, as when Laban overtakes the fleeing Jacob on the mountains of Gilead, overlooking the Jordan River: "And Jacob took a stone, and set it up for a pillar . . . and Laban said, This heap is a witness between me and you this day" (Gen. 31:45, 48). The pillar functioned both as a witness to a treaty and as a border marker between the territories of the two men. Other standing stones were understood to house the spirits of ancestors, and others still to house the spirits of gods.

The very earliest standing stones we know of in this region date to a culture called the Natufian, which flourished at the end of the Old Stone Age, about fourteen thousand years ago. These stones are worked; a human hand has engraved or otherwise modified them. But in the desert region, these worked stones quickly disappear. From then onward, the standing stones have not been sculpted, etched, or polished. Avner says they are carefully chosen for their shapes and sizes, but they remain untouched. This seems to be part of their sacred quality—we see such a concept in the Bible as well, as in God's instructions to the Israelites about the altar they are to build Him: "And if you will make me an altar of stone, you shall not build it of hewn stone, for if you lift up your tool upon it, you have defiled it" (Ex. 20:25).

"At the time when their neighbors in Egypt were sculpting their gods," Avner explains, "the people of the desert had a different approach. They chose stones whose proportions and size seemed appropriate for them and stood them up, without making any marks on them. As they saw it, the gods made humans, so how could humans possibly make them? It's not that they didn't know how to work stone. The desert produced the concept of the formless god."

We have no way of knowing what the people who erected the standing stones thought about their gods. Only the rocks, not the theology that gave them meaning, have come down to us. Presumably these people, like other ancient people and like many believers today, were trying to explain the seemingly random forces and events around them as expressions of the wills of spiritual powers. Does the refusal of the desert ancients to depict their gods mean that they believed these gods had no physical form? Perhaps not; they may simply have refused to make graven images of them. That's an innovative enough idea in an ancient world that was full of figurines and statues and paintings of divine beings. The concept that the gods have no human, animal, or plant form at all, however, is completely radical; when your gods have no faces, it becomes much more difficult to tell stories about them, or to know whether they are pleased or angry. Certainly one of the great attractions of the Christian God is that one of His persons has a face that one can draw, sculpt, and dream about.

Jews and Muslims have to deal with a God whom they can't see or depict. According to Avner, Judaism and Islam remain in this fundamental way desert religions, which trace their idea of God back to a rock in the wilderness. The Nabateans, a desert people who crisscrossed the Arava bearing spices and unguents from Arabia to Rome, also set up unformed stones, Avner says. During the Roman period, they began to establish permanent settlements, and to carve idols of their gods. Later, when many of them resumed a seminomadic lifestyle, they rejected these idols and returned to believing in gods that could not be depicted.

Avner hesitates, glancing at my *kipah*, the knitted cap that marks me as an Orthodox Jew.

"I don't know where you stand on the Bible . . . ," he queries.

"I'm religious, but I'm also a man of science. I think we've got to follow the evidence wherever it goes," I say. Avner relaxes.

"The people of the desert developed a spiritual world that influenced the rest of the world, both philosophically and ritually. They were the religious avant-garde of ancient times," he says.

Like most scholars, Avner believes that the narrative of Israel's origins offered by the Bible is not literally true. The evidence provided by archaeology, ancient written sources, and philological analysis of biblical text itself hints at a more complex story.

"I accept the Bible as an extremely important document, but one that requires critical reading," he says. Such a reading indicates that "the Israelite nation was created from among many ethnic groups, not all of whom came out of Egypt."

These ethnic groups, some of which were already residing in Canaan when other groups came out of the desert to join them, united into a single people in response to political and economic conditions. Each of these groups contributed something to a resulting common culture. According to Avner, the contribution of the desert tribes was the concept of the aniconic god—the god who cannot be depicted.

When did the doctrine that gods could not and should not be sculpted or drawn become the concept that there is a single God who has no form? The Bible's historical and prophetic books say that the belief in the one God of Israel existed for centuries alongside worship of many gods. As the prophet Elijah angrily discovered, his countrymen felt no cognitive dissonance when they worshiped both Israel's one God and the gods of their Canaanite and Phoenician neighbors. "How long will you keep hopping from one branch to another?" (I Kings 18:21), he demands of his people on Mount Carmel, in the presence of the priests of the god Ba'al. His language is precise and pointed. The verb he uses, *poschim*, also means "limp," and it alludes as well to the Israelite nation's central national ceremony, the Pesach, or Passover sacrifice. His noun carries a triple meaning: branch, thought, and a fissure in a rock. So the sentence could just as correctly be translated: "How long will you keep limping from one idea to another?"

The archaeological record indicates a transformation over time. There is tantalizing though silent evidence, for example, from Arad, which was a citadel in Judah, on a mountain overlooking the Dead Sea. The citadel contains a sanctuary built on the same plan as the Bible describes for Solomon's Temple in Jerusalem. Avner notes that three standing stones were uncovered in the temple's holy of holies, a large one flanked by two smaller ones. At some point, one of the side stones was sealed off with plaster, leaving two. At a later date, the other side stone was knocked down. The entire temple was destroyed at approximately the end of the eighth century B.C.E. At that time, according to the Bible, King Hezekiah of Judah purged his land of idol worship. Some archaeologists and historians think that the Arad temple might have been destroyed as part of that religious reformation.

Avner calls me a cab, and as we wait for it to arrive, we discuss how Eilat has grown. I get here only infrequently, every few years, I tell him, and each time the city is bigger.

"And not for the better." He shakes his head sadly.

"I just hope they don't open casinos," I say. Avner tells me that a few years ago he was a leader in the fight against the government's plans to legalize gambling here. He thinks that fight might be lost.

To drive from where I live in south Jerusalem to the Arava, you once had to take the Hebron road north and swing around the Old City. From there the road climbed the Mount of Olives, running straight through an ancient Jewish cemetery and three Arab villages and over a ridge. Then the descent began.

During the nineteen years between 1948 and 1967, when a border ran through Jerusalem, to get to the Arava you had to head counterintuitively west, circling completely around the southern bulge of the West Bank to the old British road, the Scorpion Ascent. After the Six Day War, Israelis could again use the old eastern road. But going through the Arab villages was slow and risky, so a new, wide road was built leading from the Old City north out past the new Jewish neighborhoods in East Jerusalem. From there you turned east and down.

A few years ago, in an attempt to ameliorate the capital city's horrendous traffic jams, a freeway was built through a series of rocky, scraggly valleys that had remained undeveloped as the city expanded. The Begin Freeway leads straight from the huge Jerusalem mall into the eastbound Dead Sea Road. One can now get from south Jerusalem to the Dead Sea encountering relatively little congestion, at the same time completely avoiding the Old City, Arab villages, or pretty much anything else that would signify that you are going from a city with thousands of years of history, currently claimed by two peoples and three religions, down to the lowest point on earth.

Jerusalem sits on a thick layer of alluvium, conglomerate, and limestone, which has been deposited by rivers and lakes over the last 1,500 or so thousands of years. During that time, its rocks have been pushed up like a dome, and along an intermittent ridgeline of which the Mount of Olives is a part, the rocks have been tipped over to the east. A thousand or two millennia is the most recent of times in geological terms, and the process that caused the tipping was a powerful one. The descent is raw, steep, and nearly uninterrupted. My ears begin to pop as I pass the exit to Ma'aleh Adumim, the big Jerusalem suburb perched on the eastern hills of the Judean desert. From there, the descent through brown, softly contoured hills is marked by an occasional gas station, army base, Israeli settlement, or Bedouin encampment. A fully accessorized camel stands as a sentry next to the marker that indicates sea level. A long, winding chute of a road still extends in front of me. On my bike, I could put my head down and let gravity do the work, but I'm in a car, stuck behind a large double-trailer truck. I wait for the rift to appear, but it doesn't come into sight until I am atop the last, low foothill. A vast, misty plain stretches before me. It is a hot morning after a cool week, so I can't see the mountains on the other side through the haze.

A day after descending from Jerusalem to the Arava, I find myself heading up the Scorpion Ascent. During the first half of the twentieth century, the British paved it, but it was a road for several thousand years before that. It's Monday morning, and I'm in a dusty blue Israel Antiquities Authority jeep with Tali Gini, a.k.a. Faith Erickson. Gini has been an Israeli for years and years, is married to a native, and has a son in the army, but she's still a full-figured American prairie blonde. The two hundred miles or so we clock in her jeep this Monday don't include the two-hour trip she made from her home in Moshav Kadesh Barnea, on the Egyptian border, to the Hatzeva Field School in the Arava, where I am staying. All that driving is a daily grind for her in her work as an Antiquities Authority inspector, who must cover the entire country south of Beersheba, the Negev's gateway city. She complains that it keeps her too inactive to shed the excess pounds she's gained in recent years.

Gini doesn't look Jewish, and with good reason. She grew up in rural Missouri in a family with an unusual religion. Her father, a seeker who felt no allegiance to convention, joined a sect of Christadelphians, a denomination that rejects the divinity of Christ, the immortality of the soul, and the existence of hell. His faction had a particular admiration for Jews and for the State of Israel, believing that the return of the Jews to the Holy Land was a fulfillment of Bible prophecy. So while Gini wasn't born Jewish, she grew up with something of a Jewish consciousness. When she converted in high school, her parents were supportive.

For a woman who was once part of a sect that plows the Bible for prophecies and who switched to a religion of many rules, Gini is fairly easygoing about God. While she's not observant in the Orthodox way, she's definitely a woman of religious principles. Her oldest boy, who had a difficult adolescence, was furious when she wouldn't let him bring girls home to spend the night. She agreed, after much agonizing, to her older daughter's request to go to high school in the United States but has insisted that she come home to Israel to do her army service.

"A campground," Gini said, braking the jeep and pulling over. "Probably Middle Bronze Age." That means roughly the four and a half centuries beginning in 2000 B.C.E.

It's a bald spot in the sand off to the side of the road. So what? After all, assuming the rocks that lie all around us are distributed randomly, the laws of statistics say that they won't be evenly distributed. If you lay a grid out over a large area, a lot of the squares will have an average number of rocks, more or less, but some will have lots and some will have few. An occasional square will be empty. So how can she be so sure it's a campsite?

"Look, they cleared rocks away." Gini's cell phone rings, and she lifts it to her ear. "No one wants to sleep on rocks. You can see they arranged them in a half circle around the edge. Farther from the road the area is dead. You don't find things like this," she says to me, between replies to her younger daughter, who wants to shop for school clothes in Beersheba that evening.

We commonly think of archaeologists as people who dig. They identify a mound, or the remains of a wall that signify that a building once stood there, or some historical source tells of a town or castle at a particular location. They apply for funding or university sponsorship, hire workers and advertise for student volunteers, and dig up this long-gone place of habitation hoping that it will produce interesting finds. Preferably, whatever they find will support the archaeologists' favorite theories about the period in question or, even better, falsify those of their rivals. But you can't dig everywhere, and ancient people didn't just live in their houses or workshops or fortresses. They traveled, they migrated, they moved their herds with the rains, they bought low in Arabia and sold high in Gaza. As they went, they left traces. They disturbed the random lay of the stones in the sand, scattered by flash flood and gale wind, and the stones remained fixed in their new, man-determined location for thousands of years. Discerning the human intervention in the lay of the rocks is also archaeology. Combing broad swathes of land and recording the evidence of human activity that has been preserved on its surface, rather than underground, is called an archaeological survey. Such work has become in recent decades a fruitful tool for understanding the lives of the ancients.

"I'm taking you up here to give you a feel for how people moved around the Arava," Gini said. "We think of the Arava as a north-south road because that's the principal road we see today. But in ancient times, except for in the early Islamic period, people mostly crossed the Arava from east to west."

The Scorpion Ascent's name comes from the Bible, where it marks the border between the Israelites and the foreigners around them. According to the book of Judges 1:36, "the border of the Emorites was from the Scorpion Ascent, from the rock upwards."

The Ascent is one of the region's most ancient roads. It dates back to the Chalcolithic period, the eight centuries or so that lie on the cusp between the age of stone and the age of bronze, about six thousand years ago. The demand for copper, the main element in bronze, was increasing. There was copper in the Arava. Flavius Josephus, the Jewish-Roman historian who wrote in Greek, recorded that gold was mined there as well.

To my untrained eye, the margins of the Scorpion Ascent look as nondescript as the rest of the flats the road runs through before it rises into the mountains, which the jeep is just beginning to climb. The soil is chalky, dingy, and thin. It is covered by a dense layer of mostly jagged and broken rocks, ranging from pebbles to cinderblock size. Patches of thorny, low-lying desert scrub cling to the occasional slight dip in the earth. There is an intermittent line of dusty tufts running alongside the road, measuring out what, on closer inspection, turns out to be the very slightest of channels; rain, when it falls, flows east this way. A path, relatively clear of stones, runs between the channel and the narrow asphalt on which we drive. Gini says that it more or less marks the ancient road. The road has "installations." She points to a mound of rocks along the side of the road. It's a tumulus, a burial mound. The trip from the copper mines to the coast was not an easy one. Bandits, the elements, and wild beasts all took their toll. The dead were interred alongside the road, with cairns to mark the graves.

Then there are small piles of stones, usually three in a row, sometimes on small hills by the road, sometimes perpendicular or at an angle to it, sometimes on high ridges. Gini calls them *shiniot* in Hebrew. She doesn't think they have an English name; the Hebrew would translate roughly as "tooth structures," and indeed they do look from a distance like saw teeth. Gini and her colleagues think they are road markers. I was initially skeptical, as I would expect road markers to run along the road rather than at an angle to it, but Gini points out that at an angle they are visible from far away. She attributes them to the Middle Bronze Age.

There are other piles of stones with other meanings. Gini identifies some of them as "genie mounds," created by more recent travelers, such as Bedouin nomads, seeking to ward off the evil spirits that lurk on the roadside. Each traveler who goes past adds a stone to the pile to placate the local demon. There are piles of stones that look like large, rectangular platforms and are much older than the genie mounds. The pottery scattered around dates them to four thousand years ago—the Early Bronze Age. Gini thinks that they also served as road markers—the same road in a different time. Underneath them, archaeologists have found the bones of slaughtered animals. Gini thinks these were foundation offerings, dedicating the mounds. Sometimes they were used afterward as tombs for travelers who died along the road. On one hillock by the road, she shows me that the people who made the tooth-structure road markers took stones from the adjacent rectangular platform. That shows that the road markers are younger, and also that the people marking the road had no fear of whichever god the earlier travelers had dedicated their platform to.

We hear heavy machinery, and Gini drives a few hundred yards off road to show me a great pit where bulldozers are at work.

"The whole area's turning into a phosphate mine." She shrugs her shoulders in exasperation. "It's right on the road. The Antiquities Authority is trying to stop them, but the political pressure on us is tremendous." We drive back and up the road to the remains of a small fortress that guards the Scorpion Ascent's first step up to the plateau.

"The entire Negev was transformed at the end of the third century," Gini says, segueing from the Bronze Age to the Roman Empire. "The emperor Diocletian divided his empire into four parts and assigned one to each of his three co-rulers. This area was in the part he kept for himself."

Diocletian was a divider and a builder. "Ostentation was the first principle of the new system instituted by Diocletian. The second was division. He divided the empire, the provinces, and every branch of the civil as well as military administration. He multiplied the wheels of the machine of government, and rendered its operations less rapid, but more secure," Edward Gibbon wrote of the emperor who tried to put Rome back together again by dividing it into governable units. "He had associated three colleagues in the exercise of the supreme power; and as he was convinced that the abilities of a single man were inadequate to the public defense, he considered the joint administration of four princes not as a temporary expedient, but as a fundamental law of the constitution."

With his attention focused on the east by the expanding Persia of the Sassanid dynasty, Diocletian identified the Arava and the Negev as being of vital strategic importance to Rome. He built forts, and he poured slaves—including many Christians—into the copper mines at Feynan, on the Arava's eastern side. He made the Scorpion Ascent into a real Roman road and built forts to guard it. He also placed forts along the length of the Arava and built a north-south road to connect them. This Limes Palaestinae was designed to serve as a buffer zone on the empire's border with Persia and the desert tribes in its pay.

In the summer of 1985 a water pipe burst in a field belonging to Kibbutz Yotvata, off Route 90 a few miles north of Eilat. The farmer who went to fix the pipe found that the force of the spurting water had washed a section of soil away and revealed a chunk of hewn limestone. There were letters carved on the limestone. The first line reads "*Perpetuae pace.*"

> *For perpetual peace*
> *Diocletian Augustus and*
> *Maximian Augustus and*
> *Constantius and Maximianus*
> *the most noble Caesars*
> *erected the wing with the gate*
> *by care of Priscus*
> *the governor of the province*
> *of Syria Palaestina.*

Gini flips on her left turn signal, slows down the jeep, and after a couple of southbound semitrailers go past, she turns left off Route 90 onto an earthen path that describes a short semicircle in the brush. We pick our way through some scrub bushes and thorns to the excavation site. Gini points to a weed-covered area off to the side. A Roman bathhouse. The Romans, soldiers included, never lived anywhere without a bathhouse.

Next to the farming village where Gini lives is a mound containing the remains of a Nabatean settlement that she has helped excavate. In recent weeks the Israeli army has been conducting a major sweep through the northern part of the Gaza Strip, in response to Palestinian mortar attacks on Israeli towns nearby, but Gini says that where they live is quiet. She and her husband, Moshe, moved there after leaving a kibbutz in the north. They planned to be farmers, but agriculture turned out to be unprofitable. They ended up in debt—though, Gini says, less debt than most of their neighbors—and had to look for other employment. Moshe is now an independent truck driver. He carries a lot of carbon dioxide gas from Egypt into Israel. It's cheaper to make the stuff in Egypt, Gini says, because of lower environmental standards. Gini went back to school to get a master's in archaeology, and she landed a job with the Antiquities Authority. She's recently submitted her Ph.D. thesis. The family is not getting rich, but it's doing okay. They even have a small yacht they keep in the Ashkelon marina. They use it for family cruises, and Moshe sometimes showers and sleeps on the boat when it's too late for him to make the trip home after a day of driving.

Yotvata is the site of a falling-out. Uzi Avner was eager to excavate here. He believed that the Diocletian inscription indicated that the ruins along the road were the remains of a fortress that served as a *via kaput*, a "road head," from which miles were measured—an intriguing idea because the common wisdom was that no north-south Roman road ran through the Arava.

But the Roman period is not Avner's field of expertise, and in order to get an excavation license from the Antiquities Authority, he needed an institutional sponsor. He persuaded Jodi Magness to join him. Magness brought in Gwyn Davies of Florida International University, an expert on the Roman military.

Magness, who teaches at the University of North Carolina at Chapel Hill, is a classical archaeologist with a well-deserved reputation for precision and a wide range of interests. She has written a book on Qumran, the site where the Dead Sea Scrolls were found, and another on the archaeology of the early Islamic period in the Levant. She wrote a seminal article on excavated Galilean synagogues, proving that, contrary to conventional scholarly wisdom, they are not Roman but Byzantine. Magness is also an expert on pottery—an unglamorous but critical archaeological specialty, because pottery is one of the most important means for dating finds and sites and for placing them in historical and cultural context.

Temperamentally, Magness is the opposite of Avner. While she's quite willing to row against the current when she feels the evidence demands it, she is reluctant to speculate beyond the evidence itself. It's hardly surprising that, after two years of excavating together, they decided to part ways. Magness says she offered to leave the dig, but Avner was unable to find anyone to replace her as licensee. So in the end Avner was the one who left. He continues to work at the site's margins.

Magness is characteristically cautious when talking about the finds. She tells me that she and Davies have identified three major layers of settlement. The bottom is Roman, from the late third and early fourth centuries. The second is ambiguous—it may be Byzantine, or it may belong to the early Islamic period that followed the Byzantine Christian empire. The final one, just below surface level, seems to be recent, perhaps from the Turkish Ottoman Empire. Although it was a military fort in the Roman period, it seems to have served other purposes in the other periods. Regarding her differences with Avner, Magness says only that "we have different interpretations."

Avner, typically expansive, believes there are more layers. He also tells me that the Islamic structure may well have been just one building in an entire town, then called Ghadyan, whose remains are inaccessible under the fields of the modern Kibbutz Yotvata. If so, it was at an inn here that 'Abd al-Masih, earlier known as al-Ghassani, and his monks arrived to spend the night on their journey to Ramla.

His hagiographer, writing some decades later, first portrays the young al-Ghassani, whose intentions were pure. "He was correct in worship, knowledgeable in what was his right and in what was his duty," the hagiographer instructs us. In an age when the Christian world was already divided into sects that believed all other Christians were damned, there could be no higher praise than being correct in worship. This upright and correct young Christian wished to make a pilgrimage from his home in Nagran, in southeastern Arabia, to pray in Jerusalem. Nagran's Christian community still survived under Muslim dominance. A Christian delegation from Nagran tried but failed to convince the Prophet to acknowledge Christ's divinity.

To get to Jerusalem, al-Ghassani "set out with some Muslims of the people of Nagran bent on raiding." The writer offers no explanation for his devout subject's odd choice of traveling companions. I surmise that, needing to make a long trip in territory under Muslim control, he preferred to travel with some local rowdies who he knew could protect him from their own kind along the way.

Bad choice. "On account of his association with them, they were continually beguiling him and misleading him, to the point that he went with them on the raid." Al-Ghassani's companions had every incentive to persuade the Christian boy to join their band, for he was "the best of men to shoot an arrow, the most expert of people in striking with the sword, and the most skilled in thrusting with a spear.

"Ignorance, youth, and bad companions brought him to enter Roman territory with the raiding party," the hagiographer continues. "He fought and did battle along with them. He killed, he plundered, he burned, and following their example, he engaged in everything forbidden. He prayed with them, and he became even more furious and harder of heart against the Romans than they."

In his thirteenth year of banditry, al-Ghassani decided to spend the winter in Baalbek, in the Bekaa Valley between the Lebanon and Anti-Lebanon mountains. The hagiographer does not explain why, but he says that the young man's first stop in the city was a church, where he found a priest sitting on the floor reading the Gospel out loud. The highway robber sat down next to the priest. My guess is that he was having some doubts. He hadn't meant to be a robber for so long, just for a little while, to have some fun before settling down. His thieving friends were getting on his nerves now. They were still trying to live a free and wild life, even though they were well into their thirties. The man from Nagran was getting a sneaking feeling that they would all soon look ridiculous.

"What are you reading, O priest?" asked al-Ghassani, not yet the Messiah's slave.

"I am reading the Gospel."

"Translate for me what you are reading," the young man said. He had not yet learned Greek.

"'Whoever loves mother, or father, or brother, or anything more than me, is not worthy of me,'" the priest read from Matthew.

Al-Ghassani broke down in tears, for he loved his companions at arms more than anyone else. He cried out to the priest: "Do not chide me for my weeping. I once was of the adherents of this Gospel. But today I am of its enemies. Hear my story."

When he had heard it, the priest placed a weary hand on the young man's head. "What prevents you, if you are contrite, from coming back and doing penance?"

Al-Ghassani could not look the priest in the eye. He stared at the colored stones of the floor mosaic. "I would be admitting what neither the mountains nor the lowlands will endure."

The priest cited the story of the prodigal son. Al-Ghassani prayed. The priest performed the rite of absolution and gave him communion. Thirteen years later, the pilgrim set out again for Jerusalem.

He arrived in Santa Katarina after a period of study and penance in the Holy City. When the superior passed on, the monks persuaded al-Ghassani to become their leader and gave him the name 'Abd al-Masih.

In his seventh year in this office, the Muslims raised the taxes they demanded of the monks of Mount Sinai. 'Abd al-Masih and some of his monks set out for Ramla to appeal the assessment. They traveled through the Sinai wilderness, reached the finger of sea pointing northward, passed by Aylah, and headed up the Arava in the direction of the Dead Sea. At nightfall they reached Ghadyan. There the Christians encountered several companies of Muslims returning from their pilgrimage to Mecca. A man in one of those parties suddenly seized 'Abd al-Masih with both hands.

"Are you not al-Ghassani?" the stranger asked in a loud voice.

'Abd al-Masih replied softly. "I do not know what you are saying."

The Muslim pilgrim began calling out to his fellows.

"This monk used to be with me years ago in the raiding party. He used to lead us in prayer. He was a man of the Arabs and a companion of mine, and he was hit a blow on the top of his shoulder. Examine him. If you do not discover as I have said, I am a liar."

The Muslims stripped the father superior and saw the scar. Muslims tolerated Christianity, but a man who was once a Muslim and abjured that faith had committed a capital crime. The pilgrims arrested 'Abd al-Masih and his companions and took them to Ramla. The Muslim governor suggested to the prisoner that he re-adopt the Muslim faith and so save his life. 'Abd al-Masih refused. The governor ordered him beheaded and directed that his body be thrown into a well and burned. Nine months later, 'Abd al-Masih's monks returned (never Muslims, they had committed no crime) to the well and exhumed the martyr's bones. They brought the skull of the Messiah's slave back to Santa Katarina. There, some years later, it inspired a young writer to set down the story of this man who had, like the Christ himself, been betrayed by a friend of his youth.

• • •

Tali Gini returns me to my room at the Hatzeva Field School, on the east side of Route 90, about fifteen miles south of what's left of the Dead Sea's southern basin. A relic of another Israel, back when the desert was the great frontier, the school is composed of five rows of plaster and cement cabins that wear rapidly in the desert sun and wind. This is the kind of place that idealistic, patriotic Israeli twenty-somethings went off to in the 1950s and '60s when their parents wanted them to go to law school. They came here instead to lead schoolkids on hikes and to observe the flora and fauna. A few adventurous and very foolish ones tried to sneak over the border to Petra, where the red-rock ruins of the capital of the Nabateans still stand. Some of them returned alive.

I've come here to see Amotz Zahavi, who was once one of those idealistic twenty-somethings. In keeping with the spirit of those times, he wanted to be a farmer. But a fascination with birds, a keen eye, and a sharp mind led him elsewhere. He became a professor of zoology at Tel Aviv University, and he put Israel on the map of evolutionary theory.

Zahavi is a big man. Well into his seventies, he has retained the easygoing slouch of youth. He speaks with a quiet, unimposing intensity that reminds me of the army officers I admired most, the ones who were able to get us all to work like hell without ever actually issuing an order.

His room, which is both living quarters and office, is sparsely furnished but homey. There's a couch and a coffee table and a pair of cheap armchairs. Zahavi's research assistant, Arnon, is entering data on a computer. On the ledge by the front window are several jars half filled with churning masses of maggots, juicy grubs to tempt the birds that Zahavi has studied here at Hatzeva for the last twenty-five years.

I fear that my presence is an unavoidable nuisance to Zahavi. If I were he, I'd probably be more annoyed than gratified by a visiting journalist. He is known in the scientific world as an evolutionary theorist, but his theory is an extrapolation from tens of thousands of hours of fieldwork. He watches birds called Arabian babblers in the wild and keeps detailed records of their every move and sound. It's labor-intensive, and the more distractions, the less data. In addition to me, he's got another set of visitors on the way, and I can see that his afternoon of work is lost. He's got only two days in the field each week as it is. The consolation I can offer is that I might generate some publicity for him, which might make someone with money sit up and take notice of his work.

I want to see the babblers, but Zahavi is hoping to take me and his other visitors to the field together. They haven't shown up yet, so we have time to chat. I've come to talk about evolution. The classical Darwinian view is that evolution takes place because organisms battle their competitors for survival in a hostile environment. Zahavi says that the underlying biological force is the need to communicate, both with partners and with competitors. Communication requires a sender and a receiver. But, Zahavi argues, no real communication can take place unless the recipient of the message has a way of assessing whether the information he or she is receiving is true or false.

"The message," he says, "has to contain within it an indication of the sender's reliability. The interesting question is not why a bird sings. It's why another bird bothers to listen, and to change its behavior as a result." The information they exchange allows them to gauge whether the best strategy is cooperation or war.

The puzzle that Amotz Zahavi thinks he's solved is the paradox of the peacock. Darwin's insight began with the observation that every individual organism is a bit different from its closest relatives— the other members of its species. The difference may be physical: a longer tail, a stronger leg, a lighter build. Or it may be behavioral: a louder roar, a quicker reaction time, an earlier rising time. These traits are encoded in the organism's genes, so they are inherited by its offspring. Furthermore, some of these traits will help an animal (or plant) get more food, avoid more predators, or resist more diseases. The organism's utmost, if unconscious, goal is to produce as many viable offspring as it can. The stronger and more resilient organism has better chances of surviving longer and gaining mates, and so reproducing its set of genes.

One of Darwin's greatest accomplishments was to conclude from this evidence that the traits that make an organism successful will become more common in the population, while less successful traits will eventually disappear. Over time, populations change. When enough changes take place, a new species emerges. Darwin's theory survived because it succeeded in explaining a lot more than any competing theory. However, it didn't explain everything. One of the most puzzling anomalies was the peacock.

Peacocks have gorgeous tails composed of very long, brightly colored feathers. When the peacock spreads his tail feathers—as he often does when he encounters a peahen—he makes a very impressive sight. Darwin's theory can explain how such a tail evolved: if peahens prefer to mate with the males with the largest and most impressive tails, then peacocks with genes for large tails will produce more chicks than peacocks with smaller tails, and successive generations will have larger and larger tails. But this begs the question: Why should peahens like males with huge tails?

The peacock's tail is expensive. It requires a large investment of resources—nutrition and energy—to produce and to carry. Even worse, it makes the peacock heavy and awkward. With the tail, he's less adept at evading predators and maneuvering between trees and bushes to find food. According to classical Darwinism, a peacock born with some other trait that attracts females but leaves him lighter on his feet would win out. Why hasn't such a specimen emerged?

Zahavi's theory is that it's precisely the tail's hassle that impresses the peahen. The signal is not esthetic. It's a declaration that "if I can lug around this heavy tail and *still* get away from leopards and get enough to eat, then I must be really strong and healthy— and my offspring will be as well." The tail's unwieldiness proves to the peahen that her suitor is telling the truth, because faking the signal would be too costly to be worthwhile. A weakling peacock who grows a big tail to deceive a peahen would starve or get eaten before he has a chance to mate.

"Do you have a child who likes bedtime stories?" Zahavi asks me as his white jeep bounces up and down on the dirt road leading into the Shezaf Nature Reserve, just south of the field school. His other guests were delayed, so he left Arnon to wait for them.

"My eleven-year-old daughter still insists on one every night," I reply.

"And doesn't she always ask you to tell the same story every night, even though you're sick of it?"

I grimace.

"Why do you think that is?" he asks.

I shrug my shoulders. "I guess she just wants something familiar."

"It's because she knows very well that you hate telling the same story over and over again. She wants proof that you love her. Telling her a story is a signal that you love her. But is the signal reliable? She wants to test it. How does she know that you *really* love her? If she can get you to tell her the story you don't want to tell."

The Shezaf Nature Reserve doesn't, on first sight, seem to have much nature in it. It coils along and around the lower end of the Shezaf riverbed, along a line of bluffs showing crumbly bands of conglomerate, coarse sand, and clay. Like the Arava's other rivers, the Shezaf flows only when rain up on the highlands to the west sends water cascading into the valley. But even when there is little or no rain, what water there is in the immediate vicinity flows underground into the low spots, so there is moisture to support a belt of vegetation. Perhaps *vegetation* is a misleading word, with its connotations of lush jungles; broad, shade-giving leaves; and a cacophony of chirps, shrieks, trills, buzzes, and bleats. Here in the Arava, vegetation means desert scrub and thorns. In the river channels there are trees, but they are stunted and meager compared with their cousins in wetter climes. There are acacias with acutely angled branches, brittle desert cousins of the pistachio and fig. And there's the *shezaf* itself, the jujube tree, which produces a small, reddish yellow, plumlike fruit that shrivels as it ripens in the desert extremities of heat and chill. The tallest trees around are perhaps ten feet high. Most are only slightly taller than Zahavi and I. They appear in stands, clusters of trees and brush. But even within the stands, there is plenty of walking space. Despite its appearance, however, this bare-bones wood is home to flies and beetles, snakes and turtles, mice and deer, owls and songbirds. Among the last is the only bird in Israel that lives in social groups, the Arabian babbler.

Passing a line of hothouses, we ascend a low ridge and reach a small stand of trees and bushes. Societies of babblers mark out territories they defend and patrol, Zahavi tells me. This territory he, not the birds, calls Tzeva'im. He sticks his head out the window and chirps. He waits a few seconds, cocking his ear like a puzzled parakeet, and then extinguishes the jeep's motor. He chirps again, a treble monotone, long, short, then longish. We get out of the jeep, and he calls again.

"Here they are." He points, and I see four grayish birds alight on the ground just a few feet from us. True to their Hebrew name, *zanvanim*, they have prominent tails. Their beaks are long, narrow, and elegantly curved. Two of them are about the size of my outstretched hand. The two others are perhaps an inch and a half shorter. All four hop along the ground toward us, eyeing Zahavi from one side and then the other. He tosses a grub, and one of the larger birds catches it expertly.

"These two are adopted." He speaks quietly but not in a whisper. "The couple lost their own children, and these two brothers' parents died in the next territory over. So the older birds took them in and cared for them."

"How old are the small ones?"

"About six months."

"And do they still get parental care?"

"A bit. Sometimes one of the parents will give them some food. When there's more than enough to go around. But the youngsters are close to adulthood, and adults don't feed other adults in their group just because they need food. I've seen dominant birds force a grub on a lower-ranking one who wasn't the least bit hungry."

Zahavi tosses a worm in the direction of the adult female, whom he calls by a Turkic-sounding name. Her mate darts over to intercept it, then hops to one of the younger birds and more or less stuffs it into his beak while emitting a satisfied-sounding trill.

"He wants the others to see that he's feeding his subordinate." The female hops toward Zahavi and cocks her head expectantly. This time she gets her grub. The name, he explains to me, is an acronym of the first letters of the colors of the four bands on her legs. Any new bird Zahavi and his assistants find in Shezaf gets banded, so that its behavior, status within its group, and moves among groups can be recorded.

"About fifteen percent of the time a bird will actually refuse to be fed. Why would he turn down a free snack? It's not because he's full. Often the bird that turns down his superior will happily take a piece of bread from us immediately thereafter."

Zahavi's explanation is that it's the act of giving, rather than the act of receiving, that is the key. The giver gains status, demonstrating to his subordinates, and to his mate, that he's better at getting food than other males. He gets so much that he can feed others as well. So he demonstrates that he's the strongest bird and therefore the one most worth mating with. To the other males, he's worth having as an ally. As with the peacock's tail, this handicap—forgoing food—is a demonstration of vigor.

Zahavi is certain that other ostensibly unselfish behavior he's observed among the babblers can also be explained by the handicap principle. For example, while a group of babblers feeds on the ground, one of the birds will often stand guard on a treetop. For a high school science project, Zahavi's daughter Tirza tracked who the guards were. She discovered that higher-ranking birds guard more frequently than those ranked lower in the group's pecking order. A higher-ranking bird will even often violently eject an inferior from the sentry post and take up guarding in its place. Another way the birds handicap themselves is in defensive behavior. When a babbler spots a threat—say, a snake or a raptor—he or she, in Zahavi's words, tzwicks, trills, and barks. The calls alert the rest of the group, who "mob" the attacker—they encircle it, spread their tails, and fan their wings in an attempt to deter it. A research assistant of Zahavi's noted that in groups with a single adult male, the top bird generally doesn't take part in the mob. But in groups with several adult males, the dominant one mobs longer than the others and actually interferes with lower males as they mob. In other words, when his position is secure, the top babbler doesn't waste his energy trying to impress the women. But when he has to compete for their attention, he actively seeks prestige by incurring danger—and by doing his best to make his rivals look weak.

In animal societies, according to Zahavi, combat is replaced by information collection. Just as he constantly observes the babblers, the babblers are constantly observing one another. Every move and sound is a signal that tells them about the rank, strength, and gumption of the birds around them, in their own group and in neighboring ones.

We get back into the jeep and go off road, bouncing over gullies as we drive along the riverbed. We cross invisible boundaries between babbler territories and arrive eventually in the land of Matta, home to seven individuals and a saga out of the book of Kings. Zahavi recounts it.

"The five-year-old son of the breeding couple got brave and threw his father out of the territory. He was left with his mother, two brothers, and a sister. But there was a problem, because babblers never mate with their parents, or with any other babbler who was a member of their group when they hatched. In Matta, the dominant male and the dominant female couldn't mate. But because they were dominant, no one else could, either.

"But the mother wanted to build the group's nest and lay eggs. So she went out and found her exiled husband and copulated with him. In the meantime, he had found his brother and gone into partnership with him, trying to carve out a territory. The mother built her nest in her home territory, laid an egg, and a male was born. Her son cared for the hatchling as if it were his own progeny.

"In the meantime, an unattached female tried to join the group. The son was interested because he'd then have a mate, but his mother wouldn't have it. She ejected the intruder. Then a larger, stronger female came and threw out the mother and her daughter. The mother went to live in her husband's and brother-in-law's territory, where she'd been spending time anyway, and her daughter came and joined her."

He paused to scoop up a bird who had come close to examine the bands on his feet.

"So what's going to happen now?"

"I don't know. It will probably be interesting." He sets the bird, whom he's identified as the beta male, the younger brother of the usurper, back down on the ground.

"The rule of a partnership is: that which you can do alone, do not do with a partner, because you never know when your partner might turn against you."

Many species of animals live alone, or with a single mate. Cooperation on a larger scale may develop for any number of reasons, for example, when a group can do a better job of raising the next generation than a single animal or a couple can do. In the Arava, the babblers live in a relatively resource-poor environment. A territory large enough to provide food for a brood of hatchlings and their parents is often too large to be defended adequately by a single bird or a couple. So the breeding couple needs partners. Single birds without a territory can't breed, so the fact that they can't breed as subordinate members of a group is no loss to them. Being in a group at least gives them the possibility that they might in the future gain dominance and be able to breed. Subordinates are always looking for a chance, and the dominant birds can never be secure. Each bird constantly has its eye on the others, measuring and recomputing its status within the group and its strength compared with that of its rivals. Every babbler is suspicious of his mate and incessantly seeks signs of loyalty and obeisance.

Zahavi picks up a bird to show me how you can sex a babbler from its eyes. One of the young males approaches a brother and crouches before him. Zahavi says the gesture is an invitation to play. By getting close to the ground he's taking on a handicap, Zahavi contends, showing that he's got no advantage over his brother. Even though the birds within a group retain their relative ranks fairly steadily over time, they still need to test their positions. It's not just a question of being, say, a number two. There's also the issue of what kind of number two you are.

"Take a look next time you're in Eilat and you see a Danish woman sunbathing topless. Pay attention to the guys around her, trying to get her for the evening. Why are they so aggressive at first? Sometimes a guy will jeer and even insult the woman. But then afterward he turns sweet! He's testing her. If he asks her out politely and she agrees, he doesn't know whether she seriously likes him. Maybe she'll jilt him during the course of the evening for a more attractive prospect. But if she agrees to go out with him after he's been rude to her, he can be sure that she's really impressed."

I'm intrigued to learn that this exchange of signals, this constant evaluation of the organisms around you, occurs not just within species but also between them. When a babbler on guard duty spots a hawk or an owl that poses an immediate danger to the group, the sentry emits a barking sound. All members of the group, including the sentinel, jump for cover. On the face of it, the bark is to warn the group. But some things about the warning puzzled Zahavi. First, the babbler watchdog is not just letting his comrades know about the danger. His bark can also be heard by the raptor, who may not have spotted the babblers yet. Even worse, the sentinel is attracting the predator's attention to himself, which is a poor survival strategy. Classical Darwinism would conclude that a gene for issuing loud warning calls could never spread through the population. The barkers would get eaten more often than nonbarkers, so they'd produce fewer progeny, and eventually the gene would disappear. Even more oddly, a babbler group will, immediately after hiding, fly up to a treetop and bark and call out loudly. Why give up a good hiding place and engage in activity that will attract the intruder's attention?

Zahavi argues that the babbler's bark is aimed at the raptor, not at other babblers. The message is: we've spotted you and you can't surprise us. Zahavi points out to me that such communication works only if it's in the interest of both sides. The raptor would ignore the babbler sentry's calls if they didn't provide it with useful information. But since the raptor's hunting strategy requires surprise, a barking babbler indicates that the group is not worth wasting energy on.

Such predator-prey communication is found throughout the animal world, Zahavi says. He believes that colors and markings serve a similar purpose. The markings on a tigress's face make it easy for the bull she is stalking to see where the tigress is directing her gaze. Assuming the tigress has already lost the chance to surprise the bull, she has an interest in letting the bull know he's her target. She forces the bull to maneuver, and this gives the tigress information about its strength and health—information that helps the tigress decide whether or not he's worth attacking. It's similar to a boxer's exploratory punches at an unfamiliar opponent, which test the rival's agility and moves.

Zahavi calls Arnon, who has collected the other guests and taken them to another territory. Arnon is frustrated; he hasn't been able to find the babblers. The sun is close to the horizon in the west, and Zahavi says it's the time the birds settle down to roost for the night. We get back in the jeep and drive to another part of the riverbed, where Arnon is waiting with two women and a man who are more or less Zahavi's age. Zahavi tells him to wait there and drives with me downstream. Finally we stop by a large jujube and wait.

"This is where they used to come for the night," he says quietly. "But they may have chosen a new roost."

As we wait, Zahavi tells me that he is convinced the handicap principle can also be applied to intercellular communication within organisms. As with organisms, there must be a speaker and a listener, and the speaker's message must include its own verification. The act of producing the message must come at a cost that no impostor could afford.

"Research on the handicap principle and intercellular communication could produce a breakthrough. But it's a revolutionary idea, and I'm not a biochemist. It's hard to get funding for new ideas," Zahavi says. He continues to call out to his babblers.

I admit that I'm automatically suspicious of anyone—scientist, rabbi, artist, or layman—who thinks his favorite concept can explain everything. My instinct, confirmed by forty-eight years of experience, is that the world is much too messy to have a single explanation. In fact, I'm temperamentally sympathetic with, if far from convinced by, the philosophical position which states that all grand theories are merely the products of human imagination, born of a desperate need to systemize the world around us. I doubt theories that reduce human behavior to a few simple physical and psychological instincts. The people I know, anyway, are far too unpredictable to be explained away so easily.

"What about the soldier who throws himself on a grenade to save his buddies?" I ask. "Can you really argue that he's just assuming a handicap? That he's seeking prestige? What good will the prestige be to him? What kind of communication is taking place if the organism that transmits the message isn't around to hear the answer?"

"Fundamentally, he's seeking honor, the respect of his comrades," Zahavi says.

"But he can't benefit from the respect he gains," I object.

"His willingness to throw himself on a grenade is an extension of his willingness to go to war. But no one goes to war in order to die."

"What about Palestinian suicide bombers? Isn't that exactly what they're doing?"

"You shouldn't look at the one who dies," Zahavi insists. "You have to look at the willingness to die. It's like people who try to commit suicide. What they're trying to do is send a message. They're saying that they wouldn't incur such a huge risk to their lives if their need for help wasn't desperate. But it's the help that they're after, not death. The fact that one in five hundred actually kills himself is a mistake. He took on more of a risk than he intended."

It doesn't ring true to me.

It's already dusk. Zahavi says there's no chance of finding the babblers now. We get into the jeep and head back to join Arnon and the others.

"So what advice will you give your grandsons when they have to enlist?" I wonder. "Will it be to incur danger to defend their friends, or to think first and foremost of themselves?"

Zahavi seems a bit uncomfortable with my question. "You can succeed as an altruist or as an egoist. But it's more pleasant to succeed as an altruist."

After he apologizes to his other visitors, who have come to see babblers but missed them, he takes me back to the field school. We stop by the office so I can buy his book about the handicap principle, and then, to my surprise, he invites me to his room to eat dinner with him.

It's a simple repast—a chopped tomato salad, white bread, tehinna, Bulgarian brine cheese, butter. We chat about our families, about the field school, and I venture to apply the handicap principle to Middle East diplomacy. I expect Zahavi to object, but I soon learn that here, too, he thinks the principle applies.

"Negotiation is just like any other form of communication. It's not enough to say the right thing. The message has to contain its own verification. The verification is that you incur a risk to yourself. Israel's leadership showed it was willing to incur a risk by giving up territories and persuading the population to support concessions. The Palestinian leadership said it wanted peace but wasn't willing to assume any risk. That was evidence that its message wasn't sincere."

Perhaps, I say. But it's also possible that from their point of view they were indeed incurring a risk, but it wasn't one that the Israelis perceived.

If it's not seen by the other side, then it's no risk, Zahavi insists. There's no point in talking if no one is listening.

It's about 8:00 p.m. when I leave Zahavi's apartment and head back to mine, two rows in front of his. The staff of the field school is organizing a party on the lawn near the office; a grill is glowing and someone is tuning a guitar. But other than that it is quiet and very dark. I say my evening prayers, shower, and lie down to delve into Zahavi's book.

I wake up early the next morning, just before dawn. After my morning prayers, I get on my bike and turn left out the field school gate. I head for the rising sun, toward Moshav Hatzeva, the semicooperative farming village at the end of the road. It sits on the border, perched on the fault line. The sides of the road are lined with hothouses and vegetable fields. Pickup trucks driven by Israeli farmers carry Thai workers to their labors. The Israelis wear jeans or standard blue work fatigues; most of the laborers are dressed in brightly colored pants, shirts, and with no less variegated cloth wound turbanlike around their heads. I soon leave the road for an adjacent path. It's tough riding. The soil is a powdery sand; the wind piles it up in low spots, where my wheels get stuck unless I lunge forward in anticipation. The folds of earth are rocky and rough. Ahead of me, on the horizon, are the mountains of Edom, higher, steeper, and seemingly more solid than the cliffs behind me on the Israeli side.

I continue on past the outpost along the road leading to the east. Although it's not much after dawn, off-white sand and drab scrub are already shimmering in the sunlight. The green shade of the date grove that I enter feels enchanted, a striking contrast with the light of the desert's desiccated hills. Brown lumps lie between the trees; at first I take them to be boulders, placed there for esthetic effect. But then I notice that the brown color actually comes from netting. I recall the three days I once spent working in the date groves at Kibbutz Tirat Tzvi, along the Jordan River, due north on the fault line. There we wrapped ripening dates, still on the tree, in nets to keep birds from consuming the fruit. But to the best of my memory, dates ripen earlier in the summer. Perhaps these are sucker shoots, sprouts that develop at the base of the trees' trunks. Farmers cut them off and plant them as clones of their best-producing trees. Date trees are sexed, so cultivators plant a male tree in the middle of a harem of several dozen females. They don't leave nature to run its course, however; pollen is collected and then sprayed on the female flowers.

On the other side of the magic grove, there is a pit that looks as if it has been dug to serve as a refuse dump. Beyond the pit is a sign that reads "Warning: Land Mines." It's a signal that I'm close to the border. I keep my bike on the packed-earth road. Since the rainy season hasn't yet begun, I can be pretty sure that any mines washed onto the road by last year's floods will have been detected by now. A few minutes of pedaling brings me to a tall, brick-red pillar, a border marker. A few hundred meters farther along, a lonely white guard booth stands next to a closed gate. Beyond it, I see a large Jordanian village and more hothouses. A truck trundles along the road southward to Aqaba.

When I turn to the north, I am faced with a surprise—Wadi Arava itself, a broad strip of green running between two low, drab cliff faces. This is the fault. The valley before me is a graben, an inverted plateau. Its sunken floor reminds me of the dropping floor in the Rotor, a spinning room in an amusement park my parents took me to as a child. Here, the water from the surrounding hills collects and flows toward the Dead Sea. The line is much less pronounced southward, but to the north it looks like a museum exhibit or a plaster model made by a junior high school kid.

Evolutionists have, generally, one of two temperaments. Some, like Charles Darwin, see the world as a great battleground. Every individual organism is engaged in an unending battle for life. Others emphasize equilibrium. Zahavi is one of these. His theory is based on communication, and communication is not a zero-sum game. On the contrary, where there is communication, there can be cooperation in the pursuit of common interests. If predator and prey communicate, they establish an equilibrium. If there's an equilibrium, there's also an interest in preserving the status quo.

God is unchanging, but the landscape I see before me demonstrates that the earth is in constant flux. Rivers change course as mountains get pushed up and weathered down. Seas come and go. Rain falls, then ceases; wind blows, then vanishes; hot spells follow cold ones. Life is caught in the middle, and to an Israeli living in the first years of the twenty-first century, turmoil seems to be the rule. Periods of equilibrium seem few, far off, and short-lived.

On my way back, I bike through Hatzeva. Some of the homes are the standard concrete-and-plaster structures that were put up when the village was built. But most of those have since been remodeled or reconstructed into suburban bungalows, complete with green lawns and exuberant flower beds. Many of the homes offer bed-and-breakfast accommodations, and one has turned itself into a weekend restaurant. Tractors pull wagons through the streets, full of Thai workers. One Thai woman stands on a street corner and impassively gazes at me as I pass, as if she is expecting me to ignore her presence. The moshav's natives look at me quizzically, a new face in a town where everyone knows everyone else.

On the road back to the field school, I pass an overweight guy with a bushy mustache and a white T-shirt. He puffs as he attempts a speed walk up the gentle incline. He says hello to me when I am still behind him and nothing when I pass.

"How do you know a fault line when you see it?" I ask Yoav Avni as we climb back into his jeep. Avni is a geologist, and he's my guide this Thursday.

"You're sitting on one now," he says, pointing to the drab rock face where the road has cut through a hillock.

I climb back out and stare at the cross section before me. "You mean this red stripe here?"

"That's just sand that's blown in. But look at the rock on either side."

Now that I look carefully, the uniform run of grayish white limestone I saw at first glance resolves itself. On the left of the sandy stripe, the rock is conglomerate, a concrete of pebbles and finely ground stone. On the right side, it's chalk, a finely grained stone composed of the minute shells of microscopic sea organisms.

"The conglomerate is from the Arava formation we've been talking about, two million years ago," Avni explains. "The chalk is much more ancient, about eighty million years old." In other words, it's from the late Cretaceous period, the time of the last dinosaurs, when most of this area was underwater for millions of years.

"It's called the Milhan Fault. It's a young one, one of the cracks caused by the opening of the rift."

The conglomerate is the remains of a riverbed that flowed over a former seafloor. As the Dead Sea Rift opened, it pushed the land on either side up and crumpled it, causing a web of fractures like those in a brittle, dry leaf when you press it between your hands. Here, in the hills and valleys to the west of the rift, the old seafloor was pushed up on one side; rain and wind eroded the newer rock above. After all this, two layers of rocks, of entirely different ages, lay side by side on the surface.

With his weathered face, scraggly beard, and wiry body, Avni looks like a gold prospector. But the rivers he sluices are dry, his only tool is a hammer, and the treasure he seeks is an immigrant stone with a story to tell.

"I'm a geomorphologist—a landscape geologist, not a hard-rock geologist," he tells me as we drive out of Mitzpeh Ramon, the cliffside town in the middle of the Negev Desert where he lives. He's wearing a broad-brimmed hat, and a blue flannel shirt flaps, only partially buttoned, over an old T-shirt and blue jeans.

"The landscape is a mosaic in flux. I look at the landscape and try to understand what processes caused it to look the way it does. I'm a field man, not a theoretician. I take what I see and extrapolate backward. And that's where the arguments begin."

Avni holds two jobs, one with the Geological Survey of Israel in Jerusalem and one with Ramon Science Center in his hometown. Practically, his research is supposed to lead to an understanding of the Negev region's fault structure, so that planners can take into account potential earthquake threats when they design new neighborhoods or towns. Avni, however, is motivated primarily by the pure pleasure he gets out of solving intellectual puzzles while spending a lot of time outdoors.

He's asked me to detour from my path through the rift and meet him in Mitzpeh Ramon. The rift doesn't exist in isolation, he says. As it opened, it deformed a broad belt of territory on either side. Its effects can be seen forty, even fifty miles away. The chasm broke through flowing rivers, tilted the Negev this way and that, created lakes, and reversed water flows. The landscape, Avni maintains, is aggressive and volatile.

A combat plane crosses high overhead. We're in an air force training area. Large swathes of the Negev serve this purpose. On Tuesday morning I'd set off with my bike from Hatzeva, following a trail that led west into the mountains. After about ten minutes of dodging big rocks in the path on the floodplain, I came to a sign: "Danger! Training Area. Firing Range. Entrance forbidden. Explosives." This is where combat pilots practice dropping bombs. I turned back. Avni presses on.

"Sometimes they bomb even though we've coordinated with them." He follows the plane in the sky. "Once I was out here with a colleague. We worked all morning and sat down under that tree to have lunch. I said to him, 'I bet this is the first time in the history of the world that two geomorphologists have sat under this tree.' Then we got up to continue working, and a few minutes later a plane dropped a bomb fifty yards past the tree. So doing geology here can be exciting in more ways than one."

Avni was born in Jerusalem; his father is a historian at the Hebrew University of Jerusalem, specializing in twentieth-century Latin America and Spain. Rocks started intriguing Avni during his mandatory military service in the 1980s. As a soldier, he went on several hikes in the Sinai Peninsula, which was then being transferred back to Egypt after more than a decade under Israeli control. He wondered about the mountains there: What did their color and shape say about their origins? Following the lead of his brother, he tried his hand at an archaeological dig, but it was too "inactive" for him.

"I wanted to hike," he says. "So geology seemed like the right field for me."

We stop at a lookout point on Mount Ayit, Vulture Mountain. We face the sun and the red mountains of Edom, whose western slopes are still in shadow. The sharp contours of the rocks are softened by triangular patches of effluvium, deltas of rivers that flowed when the rift was still young and its floor higher. The Arava below us slopes slightly inward, toward a central but fuzzy channel that declines gently toward the Gulf of Aqaba. The channel is dry, but water flows beneath it, and a line of scrub and trees marks the flow. Green fields dot the desert on both sides of the border.

In the distance, to our left, is the water divide, the Arava's Back. On its other side the slope is not so gentle. There, in the winter, flash floods thunder down the canyons of the Negev and Edom and sweep down to four hundred meters below sea level, where they meet what's left of the Dead Sea's southern basin. There's a plan to build a Red-Dead canal that will use the drop to produce hydropower and replenish the waters of the dying Dead Sea.

"The rift is clearly the central element in the landscape here. Intuitively, it must be the central cause of the appearance of the landscape all around it. Something has collapsed the geography on the edge of the rift inward and then pushed it up, so that there's now a difference of hundreds of meters between the Arava and the mountains on either side." Avni's statement reminds me of something I read some months ago—that the rift seems to have opened here because there was a weak line in the continental crust. The rift's course was once a line of impact between two ancient continents. The force of the impact caused the continents to fuse, but the suture, like any seam, was not quite as strong as the material around it. When the African-Arabian continent collided with Eurasia, pushing up mountains along a line running from the Alps through the Caucasus and down through Iran, the pressure may have been released along the line of an earlier union.

As Avni gets back into the jeep, eager to continue our tour, I try to imagine the look of the country before the rift opened. A single, unbroken plateau sloped down to the Mediterranean. Seven million years ago, in the same geological epoch as the rift began to open, the Strait of Gibraltar closed and the Mediterranean Sea dried up. The rivers that flowed here were massively powerful, tumbling precipitously down a steep slope. They had the power to sweep along boulders and grind rocks into sand.

We drive back north. "The rocks you see along the road here belong to what we call the Arava formation," Avni says. A formation is a layer or group of rocks that have a common origin in a common time period. The rocks I see out the window are rounded, stuck in the cement of conglomerate stone.

"The Arava formation was made by rivers, huge rivers that flowed here two million years ago. We can see from the lay of the stones that the center of the channel was over there, on top of that ridge. But today, when water flows in the Paran River, during rainy season, it doesn't flow on top of the ridge. It flows down here in a different channel. So something's wrong here. The river valley of today doesn't match the river valley from back then," he says.

Another problem. Today the Paran flows into the central Arava. But the riverbed from 2 million years ago doesn't head east. Then the river flowed north and emptied into the basin of the ancient Dead Sea. So what happened?

We cross a ridge and descend into another broad floodplain. It's the Hayyun riverbed, and of course it flows down into the central Arava.

"It's a puzzle," Avni says as he steers the jeep off the road and up a hill. We get out, and he leads me down a path to a place where the hill has been sliced away by water. He hacks at the conglomerate and pulls out a stone that is dark, ruddy, and hard, quite different from most of the rocks around it. "But with a puzzle you've got a frame. You know what puzzle pieces belong together."

"Where's the nearest place we have rocks like this?" he asks me.

I look to the east, to the plateau in Jordan, where the mountains are burnished red. "And Esau said to Jacob, give me to swallow, I pray you, of that red pottage, for I am faint; therefore was his name called Edom" (Gen. 25:30). Edom, the red land. "Over there," I say.

"That's where this stone came from," Avni says. "Two million years ago, even less, the river in this riverbed flowed the other way, from east to west."

"Wait a minute," I shoot back. "Doesn't that contradict the rift theory? The rift is older than the Arava formation. So how could a river be flowing from the other side over to here after the rift opened up?"

"Good point. But apparently the rift didn't open all at once. Remember, there can be a fault without there being a depression. A fault isn't necessarily a low place. The ancient river apparently flowed over the Arava's back, which was then still level with the mountains on the Jordanian side. It was a huge river, you can see from the extent of the Arava conglomerate all around. The wide channel and the size of the boulders mean that a lot of water flowed here on a steep gradient."

We're still following the Hayyun channel when a mountain range looms before us. Not a mountain range exactly, but a string of high points in the landscape. They rise no more than one hundred meters above the riverbed, but the sides are steep. Here, in this roadless expanse, Avni is tooling directly toward the steepest cliff face I can see. Seemingly determined to demonstrate that four-wheel drive can simulate tank treads, Avni follows a route up the hill that he claims to have taken on a previous tour. I, however, see no evidence that anything other than a goat has come up this way. The jeep nearly stalls out at one hairpin turn, but Avni does a quick save. We scale the top, and he tells me that we have mounted the Tzehiha hills. *Tzehiha* means "parched," and they certainly live up to their name.

"What do you see on the ground?" Avni quizzes me.

I see conglomerate and gravel, the same spray of rocks that one would imagine covering a dull moonscape.

"Arava formation?" I guess hesitantly.

"Correct. The same we saw below in the Hayyun plain."

"But we're on a mountain. How can an ancient river deposit stones on a mountain?"

Avni climbs out of the jeep with a geological map.

"It wasn't a mountain when the river ran here." On the map he points to two vertical black lines, two faults running roughly parallel to the Arava and to the Milhan Fault we saw earlier in the day. He points out the one in front of us, to our west, identifying it as the Tzenifim Tzihor Fault, a clear line running about three-quarters of the way down the slope of the ridge. Below it is a valley white with chalk, and beyond it a limestone mountain, Mount Tzenifim.

The hilltop riverbed and the faults together make the picture clear. We are standing on a horst, a block of rock several hundred meters deep that was pushed up as the viscous rock of the mantle beneath it flowed away from the widening rift valley. The flowing mantle stretched the crust and cracked it, creating a series of ridges and valleys that are miniature, embryonic versions of the basin and range region of Utah. It's precisely the same view one might have seen 20 or so million years ago on the spot now occupied by the Red Sea. Just as the Red Sea is a nascent ocean, so any one of these faults could one day be an arm of sea.

Since the chalk lies on top of the Arava formation, the river wash was obviously laid down before the horst went up. The chalk must therefore be younger than 2 million years old. The now-vanished lake that it testifies to thus washed these shores within the bounds of human prehistory in the Levant. We know from the tools they discarded that bands of hominins lived farther north in the rift, at Ubeidiya, just below the Sea of Galilee, as early as 1.4 million years ago. It's the age geologists call the Pleistocene, a time of advancing and receding northern sheets of ice, and apparently multiple incursions of ancestral humans out of Africa and into Eurasia.

Having climbed the parched hills, we now trundle down to the desiccated lake. The chalk formations on the surface resemble a long, thin tooth, with two roots separated by a triangular peninsula to the southwest and a slightly bulging crown pointing to the northeast. The chalk varies in color, with layers of pure white interrupted by greenish and black veins.

"The white chalk is composed mostly of the shells of snails and ostracods, tiny snail-like creatures," Avni tells me. After a few unsuccessful attempts, he succeeds in chipping a discrete ostracod out of the rock for me. It is close to microscopic and coiled like a screw. "Different species of ostracods thrive at different levels of salinity, so they can tell us what kind of water was in the lake. We can track their variations from fresh to saline and back again."

The type of animal that produced the rock also tells us that this was a constant lake; it was fed not by turgid floodwaters but rather mostly by rain, with the addition of runoff from the Tzehiha hills and springwater welling up from the lake bottom. The lake drained into Nachal Tzihor, to the north. The green veins are chalk as well, but they represent periods when the lake became very shallow and more saline. As it dried, green scum blanketed it. The black veins are solidified mud, deposited when the lake reflooded. Avni identifies two such intervals, separating three lake lives.

People lived on these shores. Avni's colleague, Hanan Ginat, found about three dozen rocks bearing the marks of human hands. There are oval limestones hewn to points, oblong phosphorites with pieces struck off both sides to create sharp edges, and lumps of flint tapered to acute angles. He classified them—as archaeologists have classified similar rocks from other sites—as awls, scrapers, and choppers. But we can only guess at how they were used by the hominins. Ginat argues that they were fashioned much the way the more numerous stone tools found at Ubeidiya were. No human bones were found in either place, but the style of tools, called Acheulean, is associated in Africa with the skeletal remains of *Homo erectus* ("upright man," a type of human that preceded *Homo sapiens*, the "thinking man" species to which we belong). The Acheulean culture appears in Africa approximately 1.8 million years ago. The artifacts at Ubeidiya date to only a fraction of a second later in geological time. They seem to represent the earliest human exodus from Africa and into the vast, uncolonized territories of Eurasia. While the Negev plain was then arid, the rift valley was lush, warm, and wet. It offered a path northward, where food and water were abundant. Ginat and Avni think that Lake Tzihor may have served as a way station for bands of hunter-gatherers coming out of what would later become ancient Egypt.

The tools are silent on the mental, psychological, and spiritual nature of the people who made up those bands. They left no art, no architecture, no standing stones. Written language would not appear for another 1,750 millennia. We have only the tools they presumably used to hunt, to butcher carcasses, to scrape hides. The same tools continued to be made in the same way from generation to generation, with only slight modifications over hundreds of thousands of years. It occurs to me that if all that were left of twenty-first-century culture after a cosmic catastrophe were a sewing needle, a pair of scissors, and a bread knife, they would tell scholars much more about us than we can ever know about the Acheuleans.

The periods when the lake dried up were local catastrophes for the bands of early humans. The earthquakes destroyed what might have been an essential resting place en route from their home continent to new frontiers. Other, more violent catastrophes also struck from time to time. The earth continued to churn beneath these inhabitants; it cracked, and chunks of it fell in or popped out at odd angles. The families who lived on these shores felt the earthquake.

"About 1.5 million years ago, another stage of deformation began," Avni says. "It lasted for a few hundred thousand years."

He shows me the evidence. The chalky strata put down by the lake lie at an angle, not horizontally, as they must have when the shells of dead mollusks settled into the mud at the bottom. The lake was upended slightly from the southeast. Water began to flow more rapidly into the Tzihor river channel, more rapidly than rainwater and springwater could replenish it. It drained for the last time and turned into another of the Negev's intermittent rivers, flowing into the Paran. Another layer of conglomerate covered the parts of the lake bed that were now parts of the river.

Today, the great rift in the Arava and its associated faults in the Negev produce a minor earthquake every few years and a major one—7 or higher on the Richter scale—about once a century.

"We're in a relatively quiet period now," Avni says as we bump down the Tzihor riverbed. "Earthquakes aren't spaced evenly over long periods. They seem to come in pulses. We don't know why. It would be useful to know. As Israel's population grows, the government will have to find room to settle people. The Negev is the only large unsettled area we have. We'll need to build cities here. So we need to know where earthquakes are more likely, which faults are active and which not. And those pulses—when will the next one begin?"

The riverbed looks like a train wreck. Department-store-size blocks of embankment have collapsed into the channel, their strata nearly perpendicular to the plane of our travel. I assume that furious floodwater flows eroded their foundations until they tipped and crashed, but Avni says it is the work of earthquakes. A crashing embankment can change the course of a river. Even though the region is monitored by seismic detectors, an earthquake can often go unrecorded. Sometimes, the only testimony comes from a geologist who sees a change in the lay of the land. Of course, the opposite can also be true.

"A few months ago the sensors picked up a fairly strong tremor along the Tzihor Fault. My boss told me to get out here and see what it did, but everything seemed to be the same."

We stop by a mushroom formation, where the flow of the river has worked away at a bottom layer of limestone. The limestone formed 80 million years ago under a sea. On top of it is river-formed conglomerate, and on top of that travertine, a limestone formed by upwelling springs. The travertine is full of little tunnels.

"They're fossils," Avni enlightened me. "Stems and branches fell into the mud around the spring and were incorporated into the stone. They rotted away, leaving their impressions in the rock."

In many years of hiking in this area, I've never succeeded in finding a fossil, but Avni quickly points out two leaf imprints in the rock. We get back in the jeep and flow, along with the river, into the channel where the Paran now runs.

The Paran is now the largest of the intermittent rivers that drains into the Arava. It originates in the highlands not far across the border in the Sinai Peninsula, collecting tributaries along the way. Its changing course can be traced by the composition of the conglomerate it has laid down. Its source hills are magmatic rock, granite, and porphyry not found anywhere else on this side of the Arava's drainage basin.

"These components can only be found in the conglomerate laid down by the Paran's main channel; tributaries originating elsewhere don't know these stones," Avni explains as we climb out of the channel to get a look from the highlands above.

The vista is sharp-edged and primeval. We stand on the edge of a cliff that plunges to the east into a narrow, deep gully. Beyond it is an angled pinnacle, lower than we are but with strata arched and twisted by some ancient convulsion. It is followed by a series of similar peaks, forming a long, narrow range. They are missing the conglomerate cover of the Arava formation. Their north sides are brown, furrowed limestone, while their south faces are smooth. Patches of Cretaceous chalk are visible low on the slopes. To their north, running straight in front of Avni and me, is a narrow valley that separates this range from another, darker one to our right.

Putting the two ridges and their median valley at our right hands, we gaze into a broad valley with a clearly defined channel running more or less down its center, coming from behind us and to our left and heading off at a diagonal to the northeast. On the other side are more chalky hills. Boulders two meters in diameter crouch in the riverbed, borne from upstream by raging waters, evidence of a time when the river's slope was steeper and its waters even faster. For while the Paran River is dry now, and is dry most of the year, its winter floods are strong enough to dissect mountains. What feels like a light rainfall in the Sinai and Negev can produce a deluge in the valley as the Paran collects rain and runoff from its huge drainage basin and sends them plunging down into the Arava. Each year's flash floods cut the valley a little bit wider, and the valley's current generous girth is evidence of millions of years of floods.

Avni points out a problem in the landscape. To the northeast, close to the edge of our field of vision, the river valley suddenly narrows and funnels itself into a tight canyon, which, he shows me on the map, drops steeply to the Arava. He calls it the Paran bottleneck. I remember hiking inside it once. The Society for the Protection of Nature in Israel has implanted metal rungs in the canyon's walls. They help hikers over the rock faces, which turn into angry falls in the wet season. How does a riverbed a quarter of a mile wide suddenly become a canyon where, in places, two people can't walk abreast?

Avni bends down to scoop up some rocks. "Look at the components of this conglomerate. Granite. Porphyry. Look at these. We call them pigeon eggs." He picks up a couple of stones that indeed resemble pigeon eggs in size and smooth, tapered shape. He breaks one open with his hammer to show me a glitter of quartz inside. "It didn't come from around here. It's from Sinai."

"So the Paran River's central channel used to run over this hill?"

"Before it was a hill." Avni points to the hill opposite us, on the other side of the channel. "You find the same stuff over there. Remember I told you that the Paran used to flow north."

"And then the Negev tilted."

"A million years ago the Negev went through a stage of energetic uplift, caused by the same forces opening up the rift in the Arava. The Paran's waters began looking for an outlet eastward. There was a wall of rock in front of the gorge. For a while, they backed up there, forming a lake, until the lake rose enough to overtop the wall and send the waters crashing into the gorge. That's why the gorge is still so narrow—it's very young."

Behind us, before we came up this hill, we saw the maroon stones of the Red River, which flowed 2 million years ago from the Edom mountains to the Mediterranean. The Paran has more or less reversed the course of that river. The Arava now rules the eastern Negev; all rain that falls here flows to the rift.

"In geological terms, though, this is all brand-new, isn't it?" I ask. "We don't see much further back than two million years ago in this landscape."

"Not true." Avni smiles and waves his hand toward the valley between the two mountain ridges that we looked at when we first came up the hill. "This fault line, for example."

I look, and as I look, the valley, suspended a good two hundred meters higher than the Paran riverbed, resolves itself into a fault line that runs roughly parallel to the river, down toward the rift.

"The Paran Fault. About six hundred million years old. Don't worry, it's inactive. It hasn't done anything for five million years."

The process that caused the crust of the earth to break here ground to a halt long before the Red Sea opened and long before the rift that created it turned due north and began to open the Gulf of Aqaba, the Arava, and the Dead Sea. Here it remains, a relic of the Paleozoic world, largely unaffected by all the cracking and lifting and spreading around it.

"All these tilting plateaus and wandering rivers and earthquakes and mountains going up and down make the phrase 'pristine landscape' sound awfully romantic," I say.

Our thoughts about the landscape we live in are typically founded on an assumption that the landscape had an original, natural state; we seek to keep or return it to that state. We seek to protect the landscape from pollution and overpopulation; we seek to preserve animal habitats or migratory routes for birds. We talk about hills and valleys, borders and rivers, as though we are bound to keep them intact. We forget that the landscape is no more constant than we; we can grasp it for an instant of time, make it ours, and do our best to wreak little damage. But the ground quakes and reminds us that the rocks move beneath us. I recall reading somewhere that if the dinosaurs had created an advanced civilization—farms, cities, roads, newspapers—65 million years of churning earth and crashing continents would nevertheless have eliminated all traces of their activity (assuming they hadn't learned to make plastic). We'd be left with only petrified bones.

"So what happens next?" I ask. "Where will the landscape be two million years from now?"

Avni's face takes on the reluctant expression that every good geologist displays when asked to predict the future on the basis of the past.

"Well, of course, I can't be sure," he says. "But I think that, fundamentally, the rift down below us wants to be an Atlantic Ocean. The Atlantic began as a spreading center in the middle of a continent that turned into an arm-shaped sea that eventually separated the Americas from Europe and Africa. I think that's what we have here. The Arava and the Jordan Valley are already more than just a transform fault, where continents slide by each other. There's a graben—the surface has collapsed downward. And grabenization occurs only where there is stretching. The crust there is thin. The mantle is welling up, heating the graben, so it expands and pushes the highlands up on either side. The mantle flows under the highlands, searching for an outlet."

"So we're going to have volcanoes in Gaza?" I ask.

"We've already had some volcanism. The Golan Heights were spewing lava when humans were living at Ubeidiya. There's a dike of magmatic rock down near the Yahel springs in the Arava. I think we'll have an ocean here."

A few weeks later I went to talk to a geologist of the other party, Aharon Horowitz, who lives not far from me in Jerusalem. His house is an aging stone structure on Hebron Road, the major artery that runs through south Jerusalem from the Old City. A mile or so down the road is a military roadblock where soldiers inspect the few Palestinian workers now allowed to leave Bethlehem and the surrounding villages to work in Israel. On the outskirts of Bethlehem is the tomb of Rachel, the biblical matriarch. Between 1948 and 1967, Horowitz's house was the second-to-the-last structure on the Hebron road, just twenty meters down from the "pillbox," an old guard post that still stands there. Horowitz and his family moved into an apartment in the southern section of the house right after the

Six Day War of 1967, when the political border between Israel and the Arabs shifted from just down the street to the Jordan Rift Valley. Then the road was narrow, and Horowitz had a large front yard. The yard has since been used to widen the road. He says that he'll soon have to vacate the house because there are plans to build a shopping center and high-rises on the lot.

Horowitz's academic home is not a geology department but rather the Institute of Archaeology at Tel Aviv University. He doesn't think that the Arava, Dead Sea, and Jordan River form a rift valley. In fact, he's the leading dissenter from the prevailing theory, which says that the valley is a tectonic boundary. Where most geologists, like Avni, see an Arabian plate moving northward along a strike-slip fault, Horowitz sees an internal drainage valley caused by folding and stretching. In 2001 he published a book to debunk the tectonic boundary theory. It's called *The Jordan Rift Valley*, and it has sold a few hundred copies. It includes a chapter by a fellow geologist, Zvi Garfunkel of Hebrew University, and another by two geophysicists. All three argue that Horowitz doesn't know what he's talking about. Horowitz asked them to contribute their views because, he says, "I think that's how science ought to be written."

Horowitz argues that there's a sociological component to the debate. In the mid-twentieth century, the theory of plate tectonics—the theory that the earth's crust is split into blocks or pieces that float on the mantle, colliding, splitting, and fusing—became the foundation story of geology, just as evolution is the foundation story of biology. Horowitz doesn't dispute plate tectonics, but he balks at using a single theory to explain everything. Geologists are scientists, but they are human beings as well. They share our human impulse toward simplification, and trying to find an ultimate explanation. But such simplification can sometimes blind us to complexities. Horowitz says that during the geological euphoria of the 1960s and '70s, when evidence supporting the plate tectonic theory accreted and pushed down the last walls of opposition, he couldn't find any journal that would publish his "heretical" idea that the Dead Sea rift might not be explainable by plate tectonics after all.

"I ask a very simple question, and so far no one has an answer," he says with a shrug. "If it's a plate boundary, where is the fault? Show me the fault, I say to them. They can't."

I suggest he take me on a field trip to see the evidence firsthand, but Horowitz demurs. "I don't go out into the field much anymore," he says. "Anyway, what would I show you? The fault that isn't there?"

I'd read parts of Horowitz's book before my day with Avni, and the dispute intrigued me. How could the same landscape, the same rocks, lead to diametrically opposed conclusions? As a layman, I am incapable of weighing the evidence myself; the data that Horowitz and Garfunkel spar over in the book is too technical for me. I have only the journalist's all-too-rough tool of confronting each side with the other's claims. I put the question to Avni during our jeep trip. "Can you show Horowitz the fault?"

"No," Avni admits. "He's right. I can't identify a continuous fault. There are places we don't see a fault. But that doesn't mean it's not there." It could be subterranean, he says, or otherwise hidden. That we can't see it can't disprove the tectonic theory. There's too much other evidence to support it. There are the copper deposits and the structural similarities to other such faults around the world.

Of course, Horowitz has explanations for all the evidence. In his view, the copper deposits, for example, mark the shoreline of an ancient ocean. Earthquakes in the region are not centered on the supposed fault line.

"I don't ignore a single one of the facts on the ground," he says.

PART II

HANGING BY A HAIR

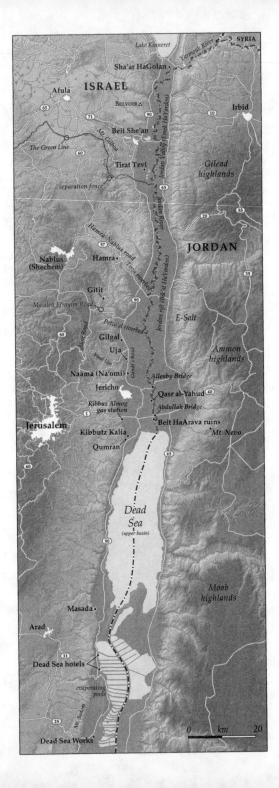

Lake Kinneret

Sha'ar HaGolan

Yarmouk River

SYRIA

ISRAEL

Afula

BELVOIR

Irbid

65

71

90

10

Mt. Gilboa

Beit She'an

The Green Line

60

Gilead highlands

Tirat Tzvi

65

separation fence

20

35

Hamra–Nablus road

57

JORDAN

Nablus (Shechem)

Hamra

35

Gitit

Tirza riverbed

Ma'aleh Efrayim Road

90

E-Salt

Petsa el riverbed

60

Gilgal

Allon Road

Ammon highlands

Uja

Wadi Uja

Gandi's Road

Naama (Na'omi)

Allenby Bridge

Jericho

40

Qasr al-Yahud

Kibbuz Almog gas station

Abdullah Bridge

1

Jerusalem

Beit HaArava ruins

Kibbutz Kalia

• *Mt. Nevo*

Qumran

60

65

Dead Sea (upper basin)

90

Moab highlands

Masada •

Arad

31

Dead Sea hotels

evaporating pools

25

Mt. Sodom

0   *km*   20

Dead Sea Works

E den, says Jerry. Utopia. A blissful heaven on earth, where there is neither illness, nor poverty, nor war.

All I can say is: He and I have different interpretations.

I stand on an embankment that rises only inches above fetid magnesium ponds on each side. To my right, in the south pool, a monster machine with a riverboat paddle wheel for a mouth grazes torpidly through greasy water, consuming ton after ton of toxic mud. Behind me is a gray steel guard post on stilts. Two women soldiers peer out from the window of a bird's nest where only a flying robot would ever lay eggs. The third member of their team, a thickset, mean-looking blonde, patrols below, giving us the evil eye. Before me is a sign that says "Frontier ahead." Around me is an assemblage of sociologists, cartographers, and geographers, scholars who study international boundaries. They have come here to examine the line that divides the blasted plain of Sodom between Israel and Jordan. In this landscape, with their pantsuits, sports jackets, and open collars, they look like a fact-finding mission to Mordor.

"We're at an international border here, so I want to give you the safety instructions the army says we have to give to everyone who comes here," Jerry declares. "In case any shooting commences, the ladies are to lie on the ground and the men are to lie on top of the ladies."

"The Arab Potash Company is over there." Jerry points to the northeast. Through the sulfur mist we can make out a factory and company town in the shadow of the mountains of Moab. "They extracted two million tons last year. We have excellent relations. Our managers and engineers sometimes go there to give them advice.

They used to come here, but since the Intifada began they stopped. Sometimes when they're short of stuff to meet a contract, we top up their stock. Not that we mention it. If you asked me, I'd deny it. But it's model of peaceful relations."

"But where is the actual border?" asks a confused Californian. She seems to be looking for a line in the sand.

"Oh, God, who gives a shit," mutters a skinny lecturer from El Paso, the only one of the group who has enough disregard for his profession to have dressed in shorts and a T-shirt.

"The Jordanians have to truck all their phosphates down to Aqaba for export." Jerry shakes his bushy gray beard and crosses his hands over his rumpled red shirt. "We've offered them the use of the conveyor belt that takes our product out to the rail line that takes it to the port at Ashdod. The belt operates at only twenty percent capacity now. But even though the two countries are officially at peace, they're afraid to get too close to us. It's all politics."

Jerry was born in Canada and has been working at the Dead Sea Works since 1968. Behind us, the factory rises up out of the flats against the wall of the Negev plateau, a huge complex of tanks, silos, catwalks, and furnaces in the midst of dunes of powder and grit. Smokestacks discharge thick billows of ash into the still air, caught between the mountains on either side of the rift valley. It looks like something Stalin built on the steppes, a glorification of the machine, industrial man pinning a recalcitrant Mother Nature to the ground.

"We have fifteen hundred employees, including three hundred in the offices in Beersheba," Jerry declares into his microphone as the bus conveys us back to the factory. "There's no attrition. People who land jobs here never want to leave. Management takes care of us, gives us everything we need."

Management's beneficences include transportation to and from work, on a staggered schedule that allows working mothers to take their children to day care before descending from the cities of the plateau to the valley below. Employees enjoy supplementary health care and generous pensions. Their children get summer jobs and college scholarships. On top of all that, the company pays an annual royalty into the Israeli government's coffers. There's an enlarged photograph of last year's check in the lecture hall where Jerry briefs us. It's for 83,550,132 new Israeli shekels—nearly $19 million.

"These smokestacks are the only ones in Israel that aren't required to have scrubbers," Gidon Bromberg tells me on the bus. Bromberg is the head of the Tel Aviv office of an organization called Friends of the Earth Middle East, and he is my green Virgil in this alkaline inferno. Bromberg says that the original concession granted to the Dead Sea Works by the British Mandatory administration, which was reaffirmed after independence by the Israeli government, exempts the plant from most environmental regulations. The heavy fuel burned by the plant's generators injects carbon dioxide, sulfur dioxide, carbon monoxide, and nitric oxide into the air. The works are also one of the world's leading producers of ozone-depleting methyl bromide.

Although it's mid-January, we're sweating. "It can get up to a hundred and twenty degrees in the shade in the summer." Jerry has the two-tone singsong of an auctioneer.

The road we ride on is made of packed salt. On either side of us, stretching as far as we can see, lie huge evaporating ponds. They used to be fed from the Dead Sea's southern half, but that dried up completely in the 1980s. So the factory laid pipe and now brings the water in from the sea's deeper, more distant, and shrinking northern basin.

"This is the largest tract of evaporating ponds in the world," Jerry tells us. "Dead Sea water is between twenty-seven and thirty-three percent salts. We pump the water up to a series of staged pools, one higher than the other. The salt settles out, and we extract it, then magnesium, and potassium. That harvester you see there is scraping up carnallite that has precipitated at the bottom of the pool. Carnallite is hydrous potassium-magnesium chloride. Potash is used for fertilizer, to help feed the world."

For those who know their Bible, the word *carnallite* may evoke the story of the two girls who escaped from Sodom and, believing that they were the last women alive in the world, plied their father with wine and slept with him. The names of their sons, Ammon and Moab, are the names of the mountains to our east.

"The healthful characteristics of the Dead Sea's waters are recognized worldwide," Jerry says. "Just north of us is a resort area with fifteen luxury hotels and spas, four thousand rooms. The health departments of Germany and Scandinavia send their psoriasis sufferers there for treatments."

Jerry's sunny confidence in nature's inexhaustible benevolence in this most forbidding of environments seems as paradoxical as a long and clumsy ornamental tail on a ground-living jungle fowl. He looks at the Dead Sea and the salt flats and sees something that most people do not. He is a true believer. But he believes in something tenuous indeed.

Jerry is not the first person to see paradise here. Some years ago archaeologists found a large hoard of copper, ivory, and woven objects in a nearby cave. Most of the crown- and scepter-shaped pieces seem to have had no practical use, so the excavators guessed that they are religious in nature. Scholars speculate that they were ornaments in a temple located somewhere along the Dead Sea coast and that priests or devotees concealed them in the cave when the temple was destroyed by an earthquake or invasion. The pieces date to the Chalcolithic period, about six thousand years ago, when Stone Age humans began using copper for special occasions.

Why did these ancient seminomads think they could find their god on the shores of a smelly lake of undrinkable water and no fish? Perhaps it was the imposing sight of the rift's mountain walls reflected on the surface. Perhaps it was the miracle of springs of fresh water welling up from the ground only a short walk from a sea ten times saltier than the Mediterranean. Perhaps they took the earth's destructive, jerky dance under the sea as evidence that great powers resided here. Or was it the very strangeness of the landscape itself? Across from the Dead Sea Works, to the west side of Route 90, is a mountain made of salt. The salt precipitated out of the waters of an ancient forerunner of the Dead Sea, was interred in the ground, and then flowed, like oil, from its point of origin to a place where it erupted through bedrock. Little grows on this amorphous lump of sodium chloride, but there's an outcropping that all the tourist buses stop at. It looks vaguely like a woman, and the guides call it "Lot's wife." This hapless biblical character turned into a pillar of salt when she gazed back at the inhospitable cities of the plain. She already missed the place that had treated her evilly. The natural sculpture is a transitory formation, about to collapse. When it does, tour guides will find another suggestive piece of rock to show their clients.

I'm as torn as Lot's wife was when it comes to paradises. My religious intuition, which lies deep in my human mind, tells me that this world inhabited by sentient beings has a purpose and a direction. The direction is upward, and we will eventually reach the time of the Messiah—sooner if we work hard at it, later if we don't. My rational instinct, inborn but hypertrophied by hard training, overlays that intuition. Based, as it is, on highly honed skepticism, I am automatically suspicious of anyone who says that the Messiah is here, or has been here, or will be arriving soon. And I know my history. Every Jew who thought the Messiah had arrived, from Rabbi Akiva through Nathan of Gaza to Emma Goldman, met with disappointment and disaster. Simple induction leads to a good working hypothesis. If someone tells you they're in paradise, they're wrong. Still, there is something wonderful about the belief in utopia, even if it hangs by a hair. I stand in awe of the fact that the only thing in this universe that can turn a gaping, pus-filled wound in the earth's crust into Eden is a human imagination.

The spa hotels that Jerry mentioned are a few miles north of the chemical plant, along the shores of an artificially maintained patch of the sea's former southern basin. As with the factory, water is piped in from the north. The hotels and the factory are at odds; as the sea dries up, thanks in part to the Dead Sea Works pumping water south, the coastline is becoming unstable. The ground is settling, and huge sinkholes open up overnight. The hotels' foundations are cracking; some are suing the works.

In the years between 1948 and 1967, Israelis saw the Dead Sea through the lens of cease-fire lines. Those borders left Israel with a narrow finger of land extending along the sea's western coastline up to about the halfway mark on the northern basin. The sea could be accessed only from the Negev highlands. The northwest coast, much more convenient geographically for vacationers from Tel Aviv and Jerusalem, was cut off. It seemed logical to build a new resort area on the southern basin, a short drive away from the city of Arad. At the same time, the country built the National Water Carrier, which piped water from Lake Kinneret in the north to the country's coastal urban centers and Negev farms. Apparently, no one put two and two together, because diverting Lake Kinneret's water meant that virtually none of it flowed, as it once had, out of the lake and into the Jordan River. The river narrowed to a trickle, and the Dead Sea began to dry up, slowly but inexorably. Its southern basin, no more than four meters deep at its deepest point, began to shrink. Soon the beachfront hotels had no beach to front on.

The common wisdom today is that building the resort area was a mistake, but the hotels stand and serve an important role in Israeli society. In order to attract customers, they cut prices. You can take a tremendous vacation in a fancy hotel for a fraction of what it would cost you in Tel Aviv or Eilat. The Dead Sea hotels offer a taste of heaven for Israel's middle class.

On the rare occasions when we have been able to get away from home for a night or two, my wife, Ilana, and I have chosen to go to Tel Aviv or Haifa. But on our twentieth wedding anniversary, two months after my trip through the rift, we bid in an Internet auction and got an amazing price at the five-star Le Meridien Plaza hotel for the two nights preceding New Year's Eve.

The paradise of a Dead Sea hotel is self-contained. You never have to leave the air-conditioned palace. True, the beach is a short walk away, but it's hot and uncomfortable. And why bother to go to the beach when there's a pool of purified, deodorized seawater in the spa? You can lie effortlessly on your back, buoyed up by dense, salt-saturated liquid, and watch your white-robed fellow guests as they arrive for their massages, mud packs, and saunas. Ilana and I went for an evening walk, but there was nothing to see except more hotels. The view was better from our eighth-floor room than from the ground.

On our second day at the hotel, after a late and huge breakfast, we decided to go out to see an important site just a short drive away. I'd deliberately excluded Masada from my October trip. Too much had been written about it; it was too loaded with associations of patriotism and self-sacrifice for my purposes. But Ilana felt like doing a short hike, and the forty-five-minute Snake Path up to the mountain redoubt was the nearest and most accessible trail.

When I first visited Masada, in the late 1970s, the facilities were romantically primitive. You took a dirt road off the two-lane highway and followed it away from the sea. Then it turned to the south and rose up a small foothill to where a run-down youth hostel and ticket booth stood. We arrived an hour before dawn, as the protocol for teenagers and young adults required, and eschewed the cable car, which in any case looked like something you'd use to launch a monkey into outer space.

My father had admonished me not to be impressed by the mountain fortress. He didn't like the fanatical suicide heroism of its legendary defenders and has to this day refused to visit the site on any of his trips to Israel. He didn't think it was a healthy model for the modern Jewish state, and he was profoundly disturbed by the fact that it was a place of pilgrimage for Israeli youth groups and that the army used it for swearing-in ceremonies. Jews, he said, have been dying for much too long. It's time for them to live.

My first trip to Israel, however, was just five years after the Yom

Kippur War, which the country came frighteningly close to losing. The idea that the country's Jews might really have to make a last, desperate stand against their enemies was not entirely paranoid. So I went, but for the twenty-seven years that followed I emulated my father and did not go back. This time I relented. I reasoned that, after a quarter century and more, the place deserved another look. Israel has in the intervening time become quite a different country from the one I'd visited as a college student. It is wealthier, more mature, and despite terrorism and its failure to obtain real peace with its neighbors, it is more or less secure; a major war that could destroy the country is unlikely. I had heard that the Masada facilities had been completely reconstructed and figured that the site would tell the old story in a new way, more in keeping with how my countrymen view themselves today.

As soon as we saw the new visitors' center before us, however, I began to feel uneasy. Perched on the peak of a foothill, it is a huge, blank-walled structure that looks more like a bulwark than a tourist spot. It's how the Crusaders might have built a castle if their architect had seen *Citizen Kane*. I parked in the lot below, and Ilana and I walked up the steep drive to the entrance.

Making our way through the crafts fair at the gates, we entered a huge, high-ceilinged hall that would befit a warriors' victory banquet. There were no tables, but there was an information desk for the confused knight, a souvenir shop where he could pass the time, and a snack bar on the deck outside where he could have a cup of coffee and strategize his conquest of the mountain above.

At the ticket window, Ilana and I went for the cheap option, hiking up, then down again. The price of the ticket entitled us to view an introductory film, so we headed to the multiplex, where a stoop-shouldered porter funneled people into cinemas according to language. We chose Hebrew and were joined by an older couple and three young people, who were probably either about to enlist in the army or had recently been discharged. There were no seats. We stood, leaning on stainless steel railings, and the film began.

The narrator was an actor I've seen several times in the theater. He told the same old story—how a group of brave Jewish rebels fled Jerusalem after the Roman legions conquered the city and razed the Holy Temple. They made their way with their women and children to the ruins of the castle that Herod had built, ironically, as a refuge in case his subjects rose up against him. When the Romans laid siege to this last outpost of freedom, the defenders chose death rather than face the slavery that awaited them as captives. The men cut the throats of their wives and children, and drew lots to decide which of them would kill the others. When the Romans broke through the defenses, they found their prospective captives already dead. Only two women and two children who had hidden behind some urns remained to tell the story.

"Slavery or death. A hard story. But tell me. What would you choose in their place?"

It's not the right question, I told Ilana as we climbed the Snake Path. The movie omitted the context. The defenders of Masada came from an extremist messianic sect that spent more of its time murdering other Jews than fighting Romans. Angry that their fellow Jews were not attacking the Roman army outside the city walls, the Masada heroes set fire to the city's food stores, reasoning that starvation would lead to a desperate battle. According to the Talmud, that act of sabotage persuaded the sage Yochanan Ben-Zakkai that all hope was lost. He concluded that Roman domination was preferable to death between the hammer of Roman legionnaires and the anvil of the messianists. Ben-Zakkai had his students spread rumors that he had died. The students wrapped him in a shroud and told the Jewish guards at the city gate that they were taking the rabbi out for burial. Once outside the walls, Ben-Zakkai walked into the Roman camp and made a deal with Titus. Titus could have Jerusalem—he could even burn the Temple. But Ben-Zakkai would be allowed to establish a school of Torah study in the southwestern town of Yavneh. My religion is the direct descendant of Ben-Zakkai's resurrected Judaism, not that of the fanatics who fled burning Jerusalem to await the Messiah at this mountaintop palace.

On our way down, Ilana and I stopped to look at the view. I pointed to an undulating set of brown-and-white stripes in the marl on the side of the hill where the visitors' center stands.

"See that? It's the Lisan formation," I told Ilana. "It's the sediment laid down by the Lisan, the salt sea that stretched from here all the way up the rift valley to Lake Kinneret from about seventy thousand years ago until about fifteen thousand years ago."

"It looks like a wave breaking on a shoreline," Ilana said.

"The curve means there was turbulence in the water. There are geologists who can look at curves like that and tell you when there was an earthquake."

Early on the Wednesday morning after I left Jerusalem for my trip through the Arava, I arrive at Qumran. I'm scheduled to meet with Naty Tzameret, who runs the site for the National Parks Authority. There's a Russian saleswoman at the souvenir stand and a dark-skinned woman at the snack counter, but no one who looks like the man in charge. Behind the turnstile, a young Arab in blue work pants and a green National Parks Authority shirt sweeps sand into a dustpan. An older companion in the same uniform sprawls in a chair against the wall. No visitors yet. A bird with bright blue plumage sits high up on a date palm frond and noisily chatters with an unseen neighbor. The morning sun shines brightly in the eastern sky, and the air is dense and wet. A mist hangs over the sea. Route 90 arches to the west below.

The ruins of the settlement, known in ancient times as Secacah, lie on a flat outcropping at the same turnoff that leads to Kibbutz Kalia. The hill is composed of the marl and clay of the Lisan formation, bedded down by that prehistoric lake. It's so soft that, with enough patience, you could dig out a cave with your bare hands.

I sit on the steps and take a newspaper out of my knapsack. The army chief of staff has declared that soldiers who refuse to participate in the evacuation of Israeli settlements in the Gaza Strip are a menace to Zionism. Rabbis at a *hesder* yeshiva, a seminary where religious boys intersperse sacred study with army service, advise their students to engage in "gray" conscientious objection—they should not declare to their commanders that they will refuse to participate in the disengagement, but they should ask for alternative assignments.

A half hour goes by, and I finally decide to ask the lady at the ticket window if Naty is around. "Of course," she says. She points at the man I assumed was an older janitor. He is still leaning his chair against the wall, taking in the sun and drinking Turkish coffee. Skinny, with a wire-brush beard and disheveled hair, his creased face testifies to long hours of work in the sun. As soon as I introduce myself, he begins to pour out a torrent of words, leaping from one idea to another with infectious enthusiasm.

"Qumran is not like other places. The question of whether the Dead Sea Scrolls were written here or somewhere else is a debate that can be examined empirically, with scholarly tools. Qumran is a wellspring of information, the only place in the world where there is real meat. Masada has a grandiose story and archaeological finds, but it's really a gimmick. The story was written by a man who wasn't there, who put a speech he wrote on the basis of Roman heroic rhetoric in the mouth of Eliezer Ben-Yair, the leader of the rebels. Here, we have archaeology and we have documents from the same site. There's nothing else comparable."

He tosses off half a dozen cases from recent controversies in biblical archaeology. Some claimed to have discovered finds that prove the literal truth of a story in the Bible, while others said that their excavations show that chapters upon chapters of the Bible's history are works of the imagination. Others still say they've found a site that proves an event but at a place or time different from that recorded by the Holy Book. He talks in fragments.

Naty mentions an excavator who claims to have proven that the Children of Israel did indeed cross the Jordan River in the days of Joshua, but farther north than the traditional site. He pauses to take one of his infrequent breaths.

"An archaeologist is not a historian. He gives the historian tools to work with. But the archaeologist is always influenced by the stories he knows. We can't understand the objects that we find without fitting them into the stories we know. Why do so many of us want to know history? So we can arrange what we find within a story. And sometimes, when the finds don't fit, we change the story."

Changing the story is the name of the game at Qumran, and Naty is a player. Probably no other site in the Levant—in the world, actually—has produced so many competing narratives. In some ways, its combination of physical finds and texts makes it a microcosm of the never-ending debate over whether Canaan's stones and the Book of Books mesh or clash. The texts in question are the Dead Sea Scrolls, the desiccated papyri and parchments discovered in the caves around the site, with the addition of some ambiguous but tantalizing references in the writings of three of the ancient world's literary giants: the Roman geographer Pliny the Elder, the Hellenistic-Jewish philosopher Philo of Alexandria, and the Jewish-Greek-Roman historian Flavius Josephus. The finds in question are the structures, fixtures, tools, and pottery found on the Qumran site itself. Unlike most archaeological sites, nearly all of Qumran has been excavated (most digs are actually samplings from patches of large sites). Despite the comprehensiveness of the original dig, however, many of its finds remain unpublished, locked away in the vaults of the Rockefeller Museum in East Jerusalem and unavailable to scholars.

Both the physical finds and the scrolls are anomalous. The Qumran site displays unusual and puzzling features: a large number of ritual baths for such a small settlement, huge quantities of animal bones, and a cemetery that seems to be much bigger than the presumed population of the site could fill. Many of the scrolls are standard—copies of books of the Bible—but many others seem to represent a community who believed that they, a small number of sons of light, were engaged in an apocalyptic battle with all the rest of humankind, especially those fellow Jews who did not accept their doctrines. They believed that a great battle between good and evil was imminent, and that they themselves were destined for paradise. The major pivot around which the scholarly and popular debate over the site turns is whether the scrolls were written by or connected to the people who lived on this hill.

Since the people who wrote the scrolls believed that the world was about to end, they did not write with posterity in mind. Since they believed that they possessed knowledge that only select initiates should understand, they did not write lucidly. In the absence of a clear and reliable statement of fact, scholars can only try to piece together the physical remains and the texts of the scrolls. The evidence, like the scrolls themselves, is composed of disconnected fragments, leaving enough ambiguity to fuel fertile imaginations and theoretical passions.

Both were on display at a Qumran conference held at Hebrew University in the spring of 2005, just a few months after my visit to the site. This was a low-key, small-scale academic gathering addressing the laboratory analysis of material and organic evidence from Qumran and the scrolls. The papers presented had titles like "Synchrotron Radiation X-Ray Microbeam Fibre Diffraction Identification of Textiles from Khirbet Qumran Caves" and "Fungi and Cultural Heritage." All offered insights beyond the comprehension of even an educated layman. But the organizers were not content to confine their conference to technical details. Like pretty much everyone else who works in the field, many of them are certain that their specific, detailed findings have clinched the argument and proved correct one of the competing theories about the site.

On the gathering's last day, the conference chairman, Jan Gunneweg of Hebrew University's Institute of Archaeology, chaired a panel to discuss the big picture. After he separated a few dozen participants from their coffee and pastries and ushered them into the library of the faculty club building on the university's Givat Ram campus, the panel members took their places. It was about as small as a panel could get, since it consisted only of Gunneweg himself and another archaeologist, Yizhar Hirschfeld. The apology Gunneweg offered illustrates just how charged the subject is.

"Yitzhak Magen said he's not coming. Magen Broshi turned down my invitation. Then Jodi Magness said that she won't come if Broshi or Hanan Eshel doesn't come because she doesn't want to be the only one defending the view that Qumran was a sectarian community."

Hirschfeld rejects the sectarian thesis. He contends that the scrolls have nothing to do with the people who lived at Qumran. None of them were found at the site itself, he notes, and none mention the settlement.

"It makes more sense to think that they originated in Jerusalem and that they were brought to the caves sometime after 66 and before the summer of 68, while Vespasian was conquering the area," Hirschfeld maintains. He rattles off some evidence—pottery, wooden vessels, the cemetery. Qumran, he says, was the estate of a wealthy Jerusalem landowner whose caretakers, workers, and slaves lived here and cultivated date farms and balsam at the Ein Fesha springs below.

"I suggest that it's much more logical that the owner belonged to the upper class in Jerusalem and that the people here were workers and slaves. The owner obviously didn't live here because there is no mausoleum on the site. Perhaps he was close to the Sadducees. Perhaps he gave orders to his workers to help the librarians who arrived from Jerusalem hide the scrolls."

It's all perhaps. Gunneweg takes off his glasses for emphasis. "No one can say Essenes lived there until we find an inscription that reads 'I'm Joe the Essene and I live in Qumran.'" There is no such inscription, and there probably couldn't be. The Essenes may well not have called themselves Essenes, and the people who lived at Qumran certainly didn't call it by that name.

Jodi Magness, by contrast, thinks the evidence clearly supports the claim that the scrolls were produced by a group of sectarian Jews living in a kind of commune at Qumran. This is not the kind of statement that makes for newspaper headlines or sound bites on CNN, because it simply confirms the conclusion reached by the first excavators of the site in the early 1950s.

I met Magness at the Shrine of the Book, on the grounds of the Israel Museum in Jerusalem, three months before my trip. The shrine, a turbanlike structure sitting in the midst of the museum gardens, is meant to represent the lid of one of the jars that held the scrolls when they were found. It is cooled from the outside by sprinklers. The entrance from the bright summer light into the dim, cool interior reproduces the transition from the Qumran settlement into the caves in which the scrolls were concealed.

More than one scholar has told me that Magness is one of the most accomplished, thorough, and professional archaeologists working today. Her field of expertise, the classical Roman period in Palestine, has assured that her name keeps popping up in connection with sites I visit on my trip. Although she never excavated at Qumran, she has authored a book about it that is certainly the most level-headed and best-written account of the site's archaeology available. It renders the technical evidence comprehensible to the lay reader.

The museum's very title of "shrine" says a lot about the way official Israeli culture wants the public to view the scrolls and the Qumran community. The discovery, rescue, and decoding of the Qumran scrolls are part of Israeli legend, much like Yigael Yadin's excavation of Masada. In this case, one of the heroes is Yadin's father, Eleazar Lippa Sukenik. On the very day the United Nations resolved to create a Jewish state, he purchased three of the first seven scrolls to be discovered. The seller was a cobbler and antiquities dealer in Bethlehem who had bought them for pennies from the

Bedouin who found them. The cobbler sold the other four to a Syrian Orthodox clergyman, who ultimately took them to the United States and put them up for sale by placing an ad in *The Wall Street Journal*. Yadin happened to be in the United States at the time and bought the four scrolls for the State of Israel, which built the shrine to display them.

But by that time Qumran was part of the West Bank, the area between the rift valley and the cease-fire lines that the Kingdom of Jordan occupied and annexed after the war of 1948. Further scrolls were discovered in a total of eleven caves around Qumran, and the site was excavated by a team led by Roland de Vaux, a French archaeologist from the École Biblique in East Jerusalem. The rest of the scrolls and de Vaux's findings were housed in the Rockefeller Museum in East Jerusalem. In the Six Day War of 1967, Israel took control of the West Bank and East Jerusalem, including the museum. The scrolls were reunited.

Museums don't like the "evidence suggests that," "we may presume that," and "it may very well be" dictums of careful scholarship. Curators believe—quite correctly—that the public wants a clear story. The Shrine of the Book presents the sectarian hypothesis about the scrolls as if it were indisputable fact. Even though she agrees, Magness points out to me how the display irons out ambiguities.

"These lamps aren't lamps from Qumran." She points to a window display of several of the objects. "They're from the same family but not from the actual site."

The Shrine of the Book emphasizes that Qumran, once outside Israel in the Jordanian-ruled West Bank, was originally inhabited by Jews and that the Jews who lived there wrote the scrolls. The emphasis is on their Jewishness, not on the fact that they were Jews of a peculiar sort, who were actually considered heretics by many of their fellow Jews because of their eschatological beliefs and deviant practices.

At Qumran itself, it's a different story, but it's in transition.

"Until the Intifada," Naty tells me, "we had 450,000 visitors a year, and ninety-seven percent of them were Christians. Hardly any Israelis. Last year we had 80,000 visitors, twelve percent of them Israelis. With foreign tourism down so much, I'm convinced the only way to keep the site going is to gear it more toward Israeli tourism."

Why are the ruins of a Jewish cult settlement so interesting for Christians? One of the peculiarities of the Qumran site is its large number of ritual baths. Ritual immersion is familiar to Christians in the form of baptism. While most modern Christians and Jews think of baptism as a peculiarly Christian practice, ritual immersion is in fact part of Jewish ritual as well. In modern observant Judaism it is mandatory only in specific cases, and mostly for women. In the time of the Qumran settlement—the first century B.C.E. and the first century C.E.—the laws of ritual purity were observed strictly, at times to the point of obsession. Rabbinical texts of the time inform us that especially devout Jews sought to maintain a level of purity equivalent to that required of the priests serving in the Temple in Jerusalem. A person could be rendered impure quite easily, too: not only did direct contact with a dead insect or a dead human body render one impure, but contact with someone who had touched someone who had touched a dead insect or dead human being, or contact with a liquid that had been in contact with them, could also render one impure. One privilege of full initiates into the Qumran sect was participating in communal meals, and immersion was required before each one. Initiates thus had to immerse themselves several times a day.

According to the New Testament, John the Baptist conducted public ritual immersions in Judea at about the same time that Qumran's residents were immersing themselves in their baths by the Dead Sea. Jesus, according to tradition, was baptized by John near Jericho. So it has been very tempting for Christians to identify Qumran with John the Baptist and with Christ. John is not mentioned in the Dead Sea Scrolls, and his practices seem to have differed from those at Qumran, but phrases such as "John and Jesus may have been part of the Qumran community at some time" are enough of a draw for Christian tourists.

The sun bakes my back as Naty shows me one of the baths. Several broad steps lead into a now-empty rectangular pool. As sweat drips down my neck, I find myself wishing it were full. The settlement's water was distributed to the baths by earthen channels. A ground-based aqueduct channeled water into the settlement from the Qumran riverbed above. The river is dry throughout the year; it flows only during occasional flash floods in the winter. These are brief and violent, and huge quantities of water flow from the plateau down into the Dead Sea. One flood was enough to fill all of Qumran's cisterns and provide the community with bathing and drinking water for a year.

We gaze out at cave number 4, dug into the marl of the cliff face on the other side of a gorge that separates two of the hill's fingers. Magness believes that the Qumran sectarians lived in the caves and that the site itself was used only for communal activities, such as eating, study, and the composition of the scrolls.

"We get a lot of crazy amateur archaeologists here," Naty says. "They're convinced they know where to find the hidden scroll that will prove Jesus lived here, or in which cave the priests hid the sacred objects from the Temple when the Romans besieged Jerusalem. One guy last year came with an elaborate proof of why there were more scrolls buried here, on this spot where we stand. He dug fifteen meters down and didn't find anything. He wanted to keep going. I told him that enough was enough."

"Why did the Essenes, or whoever they were, come to this dry, exposed hilltop to build their community?" I ask him. "Were they spiritually inspired by the view of the mountains and the sea? Were they seeking isolation? Were they fleeing persecution?"

"We can't know." Naty shrugs and leads me back toward his green-and-white National Parks Authority jeep. "Maybe the property was cheap. Who else would have wanted it?"

•   •   •

The jeep vibrates, bounces, and sputters. In the passenger's seat, my bad left ear, rendered nearly useless eight years ago by a major illness, is pointing in Naty's direction. He speaks with passionate conviction against a recent archaeological interpretation of a large collection of date pits found in one of Qumran's corners, but I can't really make out what he's saying. After we turn north on Route 90, I notice that the road lies far from the Dead Sea shoreline and has a different shape. When I first came to Israel in the 1970s, the sea nearly reached the road. Now the road retains the image of the former shore the way a fossil retains the image of a Mesozoic mollusk.

The road bears right around the northern end of the sea until it reaches a T intersection where a lonely gas station stands. There's a barrier blocking the continuation of the road to the east, declaring it a closed military area. A battered, chipped, and faded two-story pink-and-blue plaster-and-cement building stands a bit beyond the roadblock, along with similar but lower structures. Naty takes the left turn, and we drive north for a few minutes before the road veers west, in the direction of Jericho and Jerusalem.

Until the mid-1990s, Route 90 here headed away from the center of the rift valley. It then turned right at the foot of the mountains on the west side of the rift, ran straight through Jericho, and continued north toward Lake Kinneret. In the late 1990s, when Israel handed Jericho over to the Palestinian Authority as part of the Oslo agreements, Route 90 changed course. Jericho was always a relatively calm and peaceful city, but Israel wanted to allow citizens traveling north through the Jordan River Valley to circumvent it. The old military patrol road to the east, just a kilometer from the river channel, was widened and repaved. Israelis were then able to travel the length of the rift valley without encountering its only Arab city.

When tensions increased and the second Palestinian uprising broke out at the end of 2000, Israeli citizens were forbidden to enter Jericho, even if they so chose. The new Route 90 became the only way north. A year later, Minister of Tourism Rechav'am Ze'evi was assassinated, and the road was named after him. As general of the central military command, Ze'evi had been responsible for security along Israel's eastern border in the 1960s and had encouraged Israeli settlement of the valley. He was a demanding, no-nonsense officer who was a stickler for maintaining nationalist morale. He banned the army entertainment corps from singing a hit called "The Peace Song" to his soldiers on the grounds that it would diminish their motivation to fight. He later entered politics as leader of an extremist party that advocated transferring the Palestinian population of the West Bank and Gaza Strip to Arab countries or elsewhere. His dark complexion and slight frame seem to have been the source of his unlikely nickname, so this stretch of Route 90 is called "Gandi's Road." I don't know whether the omission of the *h* is deliberate, but the reference is as clear as it is incongruous.

Nearly a decade old, the road still looks brand-new, perhaps because it isn't used often. As the peace process soured, Israelis got more and more uneasy about traveling through the West Bank. The vicious Palestinian suicide bombings and drive-by shootings that followed the collapse of the Camp David conference in the summer of 2000, followed by the outbreak of a full-fledged uprising, slowed Israeli traffic to a trickle. Most people traveling from Jerusalem or Tel Aviv to the north now take the Trans-Israel Highway, a toll road that opened nearly simultaneously with the Intifada. You have to pay, but it's fast, convenient, and far away from the rift.

Naty has Route 90 to himself, and he takes full advantage of it. The road is mostly straight, but the jeep's path is complex—Naty talks and steers while simultaneously fielding phone calls from his staff at Qumran. He pulls up abruptly at a turnoff marked by a brown sign. It points east and says "Qasr al-Yahud." Dialing the number of the local army command, he proceeds down the road to a point where it is crossed by a fence.

An army patrol vehicle idles on the far side. A first lieutenant in his late twenties is in the process of unlocking the gate. He and his soldiers are reservists, as is evident from their age and less than sparkling appearance. The driver's olive-complexioned face bears the creases of a very recent nap. Three soldiers sit in the vehicle's open rear. One hasn't shaved this morning and is wearing the kipah of an Orthodox Jew. Another has tousled hair and glasses and sucks anxiously on a cigarette. The last one, his hand casually draped over a machine gun, is an Ethiopian with a baby face and a skeptical smile.

"I hope you won't be long," the officer says to Naty as we drive through the open gate. "These guys are all supposed to head out on a three-day leave as soon as this patrol is over."

"We'll be here about an hour." The lieutenant grimaces and climbs silently into his commander's seat. I am reminded of the timelessness of reserve duty. When you are in the army, there seems to be no past and no future, only an endless now.

Naty's jeep lurches forward, and we proceed down the one-lane road toward the river. The asphalt, bordered intermittently by cultivated date palms, looks like a long finger of modern civilization cutting through an ancient plain. It runs, straight as a compass needle, between sandy mesas carved by floods of centuries past. On the right, a dark stone castle emerges from the sand, its windows shuttered and its walls pocked by bullets and erosion.

"Qasr al-Yahud. What does that mean?" Naty asks himself, for my benefit. "In Arabic, *qasr* is a fortress. But that's what it means when you write it with the hard Arabic *s*. Arabic has two flavors of *s*. If you use the soft one, *qasr* means 'to cut through.' In my opinion, that's the correct root. Because this is the point traditionally identified as where Joshua and the Children of Israel crossed the Jordan into the Land of Israel. Josephus mentions it. So Qasr al-Yahud would mean 'the Jews' crossing.' And indeed in this area, close to Jericho, it is the easiest ford in the river. There are seven churches here."

As my eyes follow his arm, I see more structures beyond and to the side of the dark castle. Some have triangular, Romanesque façades, others are rotunda-shaped. Some are stately, others little more than modest stone rooms inlaid on flagstone patios. "They serve pilgrims who come to pray at the site where Jesus was baptized. John the Baptist didn't choose this sight by chance. He baptized believers here because it's where the Children of Israel entered the Promised Land. When Helena, the emperor Constantine's mother, made her famous pilgrimage to Palestine, she ordered a church built at this holy spot."

He pulls the jeep up by the dark castle. The patrol car follows obediently, circling so that the lieutenant's gaze is directed at the road and the soldiers in the back face east toward the river and west toward Jerusalem. Naty quotes from the book of Joshua: "And it came to pass, when the people removed from their tents, to pass over the Jordan, and the priests bearing the ark of the covenant before the people; and as they that bore the ark were come to the Jordan. And the feet of the priests that bore the ark were dipped into the brink of the water (for the Jordan overflows all its banks throughout the time of harvest) that the waters which came down from above stood and rose up in a heap very far from the city Adam, that is beside Zaretan, and those that came down toward the sea of the Arava, the salt sea, failed, and were cut off, and the people passed over opposite Jericho."

Naty comments: "Of course, the archaeologists debate about where exactly it happened, or whether it happened at all. This is the Monastery of St. John the Baptist. It was inhabited until 1969, when the army discovered that Palestinian guerrillas who infiltrated across the border were holed up here. It's been empty since then."

We walk to the river as Naty identifies the capellas for me—one Coptic, one Armenian, and so on. No Protestant churches, I notice. "No, they don't like the ceremonies here," Naty says. "Too much pageant and ritual. But if you ask me, it's an impressive sight when processions come in from the desert. You have to admit that it's inspiring, the wilderness all around, untamed and unpeopled, and here a place of worship."

The road leads to a broad, stone-paved platform overlooking the river. The trees the Parks Authority has planted around the clearing look a bit scraggly, and there are streaks of silt in some places. "We had it set up very nicely, but then the river flooded in the winter of 2002. Hard to believe, but the river came up almost to the church, and washed away everything it could tear loose."

A stone staircase leads from the platform to a smaller landing ten meters below, at the level of the river itself. Here, at the end of its course, the Jordan is barely larger than the creek that ran through the park near my house when I was a boy. Its water is turgid and sluggish. On the Jordanian side there is a small church, clean and obviously quite new, as if it has just been unwrapped and taken out of its box. Six pilgrims wait below it—an elderly woman in a black dress; a broad-shouldered, sandy-haired German and his young wife; a young Slav and his girlfriend, with knapsacks at their feet; and a young, dark-haired woman with a Bible in her hand. A middle-aged, saggy-cheeked Dominican in white vestments with a bored expression on his face immerses a seventh devotee in the water, muttering a benediction I cannot hear.

"We want to develop this as a major tourist site, but the army still keeps it a closed military area, and you have to coordinate every entry with them," Naty tells me. "And since the violence began, there aren't many Christians coming from overseas. Then a flood washed half the site away. The Jordanians realized that they had an opportunity and built themselves a church, and now pilgrims who want to be baptized at this site come through Jordan and do it on the other side."

When I visited the site later with the border scholars, I discover that the murky water contains more than the pilgrims probably suspected.

"Over the last fifty years there's been a tenfold reduction in the amount of water that enters the river," Avner Vengosh, an Israeli hydrologist, tells us. "Most of the river's water used to come from

Lake Kinneret and from the Yarmouk River, which flows in from the Syrian highlands just south of the lake. But Israel, Syria, and Jordan have diverted nearly all that for irrigation and domestic consumption."

He has conducted a long-term survey of the river's water for the Israel-Palestine-Jordan water cooperation project, in which Israeli and Arab scientists work together. Qasr al-Yahud is one of their sampling sites. "From the samples we take we can see, through chemical and isotope analysis, the fingerprint of what water is entering the river," he explains. He holds up a sheaf of graphs.

"What we see is that a lot of it is groundwater coming from underground springs, some fresh, some saline. The only other source of water is wastewater. In the summer, when there is less subsurface flow of water, sewage is the only thing that keeps the river from drying up."

He pauses for emphasis. "The peace treaty between Israel and Jordan commits both to halt the flow of wastewater into the river. But our conclusion is that the wastewater shouldn't be removed. If it's taken out, there will be no Jordan River."

The short El Paso professor with the glasses pricks up his ears and gestures toward the priest in the water. "You mean these guys are dunking themselves in shit?"

"It's not as bad as it sounds. The river cleans itself. The ammonia in the sewage oxidizes into nitrates that precipitate out of the water."

"Still, shit. Not very spiritual, is it?"

When we come up from the river, Naty and I find the lieutenant pacing alongside the patrol car with an impatient expression on his face.

"Done?" he asks sharply.

"We're done here. I want to show him the bridge."

The soldiers sigh collectively and put their helmets back on. Naty leads me to his jeep, and we set off on another narrow road running parallel to the river. In a few minutes we reach the sand-swept remains of an east-west highway. Naty parks his jeep, and the patrol car pulls up alongside. I follow Naty and scramble down the bank to the river. Above us are the twisted beams and spans of the bridge that once bore the highway's river crossing.

"This highway is a continuation of Route 1, the Tel Aviv–Jerusalem highway, and leads to Amman, the Jordanian capital. Before 1948 it was a packed dirt road that served the Dead Sea Works. After the war, when Jordan annexed the West Bank, they paved the road and built this bridge, which they named the Abdullah Bridge, after the current King Abdullah's great-grandfather. It was the king's road. He used it when he traveled to Jerusalem." Abdullah was assassinated in Jerusalem in 1951, after praying in the al-Aqsa Mosque.

"Israeli forces blew up the bridge, along with other bridges across the river, during the Six Day War in 1967. But they left the ruins here."

The contorted wreck is majestic and eloquent. The river, which we Israelis are accustomed to think of as a natural border, is easily bridged. For nineteen years a country spanned the river. That was no anomaly—most of the region's rulers throughout history, from Solomon through Herod to Saladin and 'Abd al-Hamid the Turkish sultan, have united parts or all of the west and east banks. It's the current situation, in which the river and the Arava are a border as well as a fault line, that is unusual.

"Just one more stop," Naty reassures the impatient soldiers as we clamber back up the bank.

In two minutes we're at the edge of the sea, alongside the dilapidated plaster buildings we saw beyond the barrier on our trip out from Qumran. Naty brakes his jeep. The two-story building looks even more forlorn up close than it does at a distance. It reminds me of the hovels and shells of the Ein al-Sultan refugee camp north of Jericho, that bleak and decaying monument to a national tragedy that I no longer see on my trips north because the new Route 90 detours around it.

"I've always wondered what that was," I say to Naty.

He has a dreamy look on his face. "It's the Lido."

"The Lido?"

"The original Dead Sea hotel. There was a café here, too, on the water." He points at the western highlands, streaked with gold by the afternoon sun. "In British times, Jerusalem's high society used to come here and lunch on the lakeshore." On my way back to Jerusalem on Friday, I will stop at Kibbutz Almog, on the plain just before Route 1 ascends the mountains, to see photographs from that era. One in particular will captivate me: two women in broad-brimmed hats hold teacups and chat with a well-fed man in a dark suit on the Lido patio, a serene sea behind them.

"And these other buildings?"

"The original Dead Sea Works. And over there, what's left of the old Kibbutz Beit HaArava."

As Naty tells the story, I realize that these are the ruins of a Camelot, battered testimony to an infant utopia where classes and peoples lived for one brief moment in harmony. Or so they thought.

In 1911, when the Turks still ruled and the northern shore of the Dead Sea was barren and unpeopled, a Siberian Jew named Moshe Novomeysky stood on this spot. He was Siberian because his father's father and one of his mother's grandfathers had been exiled to the remote city of Barguzin as punishment for their revolutionary activities. Novomeysky studied mining engineering and developed a method for extracting chemical salts from Siberia's briny terminal lakes. Despite his revolutionary ancestry, he displayed a Republican's talent for enterprise. By 1905, when he was jailed for seven months for his part in that year's abortive revolution against the czar, he had already built two chemical factories. After being released from jail by an imperial amnesty, and faced with the pogroms of that year, he fled to Berlin on a false passport. There he happened on a geological report on the Dead Sea's water and realized that the processes he had developed for mineral extraction could be adapted for it. During his first trip to Palestine, in 1911, he identified the northern shore as a promising spot for a factory. There was freshwater nearby in the Jordan, oil shales in the western mountains to provide fuel, and a labor force in Jericho and Jerusalem. He returned to Siberia, endured the World War and the Revolution, and moved to Palestine, now under British rule, in 1920.

After a long and unrelenting lobbying effort, he managed to convince the British administration to grant him an exclusive concession to extract minerals from the Dead Sea, and a tract of land on the shore for his factories. He built two facilities, a smaller one here and a larger one on the western shore, the site of the modern Dead Sea Works.

At the time Novomeysky moved from the USSR to Palestine, other Jewish revolutionaries were doing the same. These socialist Zionists viewed themselves as both Jewish patriots and loyal members of the international proletarian movement. One division of that larger movement had an ideology that cut through today's boundaries between left and right, nationalist and internationalist. Inspired by a white-bearded guru, they were territorial maximalists—they utterly rejected the idea of dividing Palestine into two states, Jewish and Arab. They believed that all of the Land of Israel was the inheritance of the Jewish people. But they also rejected the very idea of the state. They dreamed of an anarchist utopia made up of autonomous Jewish and Arab agricultural communes, united in Marxist brotherhood. With utter sincerity, they believed that Palestine's Arabs would welcome Jewish settlement in their country. The Arabs, they were convinced, would realize (after being disabused of false consciousness) that the Zionist enterprise would raise up the oppressed Arab workers and peasants.

The nickname given to members of the youth movement associated with this ideology was *chugistim*. Like all socialist-Zionist youth movements, their highest purpose was the establishment of Jewish agricultural settlements. But since the chugistim were ideologically opposed to removing Arab farmers from their land, even if the land was purchased, they took it upon themselves to settle only uninhabited areas—that is, the most barren, forbidding, and inhospitable parts of the country. One of the *chugist* rituals was a hike around the Dead Sea—recapitulated by their modern youth movement heirs, after a long hiatus, just a few months before my trip. Their geographical symbol was Mount Nevo, the peak on the other side of the Jordan where Moses the Prophet saw the Promised Land he would never enter. Mount Nevo was also a major symbol of Rachel the Poetess, whose three slim volumes were, along with a Bible, stuffed into the pockets of every chugist. The mountain is a metaphor for a life spent gazing upon a heart's desire that cannot, or should not, be reached.

With a nearly mystical attachment to the driest deserts and most malarial marshes, the chugistim proclaimed it a patriotic duty to settle every spot that no one else would touch. Novomeysky's project offered them everything they wanted: hard labor, shoulder to shoulder with indigent Arabs; intolerable heat; and a barren desert on the shore of an undrinkable lake. Teams of them signed up to work at Novomeysky's factories. On March 21, 1940, they laid the cornerstone for a kibbutz at the northern tip of the Dead Sea, adjacent to the potash works, and planted their first date trees. They had originally planned to call the kibbutz Nevo, but they finally adopted the name Beit HaArava, a place name from the Bible's book of Joshua, cited as lying on the Israelite frontier north of the Dead Sea. A woman among the founders later recalled that some of the settlers fled on the first night there, terrorized by the sight of naked white mountains in the cold light of a gibbous moon. As for the rest, their founding declaration proclaimed: "For we have set out to war against harsh and mighty Nature, and the fury of war will rise among us, and it was good in our eyes when we said: the bread that we will bring forth from our land with heavy labor will be sweet, because we have won it with our sweat, because we share it among us and in a world of theft we have in justice and in uprightness eaten bread."

One of the photos at the Kibbutz Almog exhibition shows a row of pavilion tents on the dead seashore, the first home of the settlers. In another, three women hold four of the kibbutz's first babies. There's a picture of America's best-known radical farmer, Vice President Henry Wallace, paying a visit to Beit HaArava, and one of muscular, bare-chested young men spreading a net in one of the kibbutz's fishponds. Life imitates the fantasy of a Soviet socialist realist poster in these images. The most improbable one shows a young man and a young woman stretched out on their bellies, face-to-face, on the kibbutz lawn, with open books in front of them.

A lawn? On the salt flats?

"This was their Garden of Eden?" I ask Naty, gazing up at the decaying Lido.

"I've talked to the old-timers who were forced to leave. They bear a scar as deep as a canyon. They've never recovered."

On November 29, 1947, the United Nations voted to partition Palestine between the Arabs and the Jews. According to the plan, the site occupied by Beit HaArava was to become part of the Arab state. By that time, the kibbutz was home to 120 adult members of the commune, 46 children, and 40 Jewish soldiers who were stationed there. There were vegetable gardens, a rose garden, bananas, dates, a barn, a chicken run, and fishponds. There were five graves in the kibbutz cemetery. The War of Independence began. In April 1948, the road from Beit HaArava to Jerusalem was cut off by the Arab Legion. Soon thereafter the Israeli government ordered the settlers to abandon the kibbutz. They refused at first, but under pressure, they allowed the women and children to be evacuated. When the legion finally invaded, the rest left by boat for the southern factory.

A solitary bird, perhaps a lost crane or cormorant, beats its wings high above us.

"There was no battle here, but it was over. The Jordanians razed the kibbutz. Years later the settlers learned that Novomeysky had reached a secret deal with King Abdullah. In exchange for a quiet evacuation, they'd let him keep his interest in the northern factory. The settlers felt betrayed. If they'd known, they say, they would have fought to stay on their land. They lost their paradise, and they never found it again."

At this point, journalistic convention, my instinct for fairness, and my intellectual curiosity all demand that I bring in an Arab voice. I must turn my attention to the date-palmed boulevards of the oasis to the northwest. Jericho is the largest city in the rift between Eilat and Beit She'an, and it is much more ancient than either. In fact, it has a claim to being the oldest city in the world. At its northern end are the remains of fortifications and a tower dating back ten thousand years—well before the time humans learned to fashion metal. The wall and bastion are testimony to an organized society, one in which a strong leadership was able to mobilize citizens or slaves to carry out major construction. They also bear witness to an ancient conflict: that ancient ruler would not have gone to such trouble had his city not been often under attack, probably from nomads living in the surrounding wilderness.

Route 90 used to run straight through Jericho. Israeli buses, plying the road between Jerusalem and the Galilee, used to stop for a break at one or another of the roadside restaurants by the ancient mound. Huge nets of oranges hung from awnings over wooden tables where Israelis and tourists munched falafel or scooped up hummus with pita. Jericho was always the quietest of the Palestinian cities, but that hardly means there was no tension. Most Israelis were happy when the bus drivers switched their patronage to the restaurant and gas station that Israeli settlers set up to the north. In the late 1980s, during the first Intifada, terrorists bombed a bus at the city limits, and a Jewish woman died trying unsuccessfully to save her three young children.

I last went through the city in 1994, in uniform and with a rifle. I and three others from my reserve infantry company drove through on our way to rejoin our unit on Mount Hermon. Earlier that year the city had been handed over to the newly constituted Palestinian Authority, after Prime Minister Yitzhak Rabin and PLO chief Yasir Arafat signed the Oslo accords. Palestinian policemen directed traffic in full military uniform, and paired Palestinian and Israeli checkpoints stood at each end.

At the snack bar of the Kibbutz Almog gas station, I try to strike up a conversation with a couple of the Arab workers. They have special permits that allow them to cross the frontier from the silent city, and that's about as much as they are willing to tell me. They are most likely afraid of losing their privilege, either by angering their Jewish employers or by calling too much attention to themselves among their unemployed and discontented peers at home. For better or worse, I will have no Palestinian voice where one is needed. That, perhaps, is more eloquent than anything I might hear in Jericho.

Between the salt sea and the freshwater lake of the Galilee, ancient history fades into the background and modern Israel comes to the fore. Like any grand political project, the modern Jewish state was founded on a dream, and like that of any successful political project, its success has a lot to do with the hard work and pragmatism of its leaders. The dreamers at Beit HaArava believed in many incompatible and imaginary things. They were Jewish nationalists, but they believed in the universal unity of the international proletariat. They rejected as primitive superstition the religion of their forefathers, but they accepted the premise that the Jewish people must live in the land promised to them by that very same tradition. That land, they insisted, spanned the Jordan, as if they could by an act of will close the rift that titanic natural forces had created. The Jews who now live north of the Dead Sea are much more practical. But their impending loss—if that is what it comes to—is no less real.

"No soldier will have to drag me. No soldier will, God forbid, have to feel awful if I have to get up and leave my home. But I will not hand my home over easily." Not just Beit HaArava but also Masada and Qumran resonate in Liora Hasson's voice. So does Gush Katif, the cluster of Israeli settlements in the Gaza Strip that the government has designated for evacuation. But her words set her and her neighbors in the Jordan rift apart from the settlers in the strip. According to the newspaper that Sunday morning, the settlers in Gush Katif say they won't leave, even if Israel's democratically elected government decides that they must. They'll try to persuade the soldiers sent to evacuate them to disobey orders. If the soldiers persist, the settlers will kick and scream while being hauled out of their homes.

Liora Hasson is a compact, energetic woman in her forties with short, copper-colored hair and matching skin. She and her family live in a well-appointed one-story home in the Jordan rift. In Israeli parlance, the southern part of the river valley, the part that is in the West Bank, is called the Jordan rift, Bik'at HaYarden in Hebrew. The name Jordan Valley is reserved for the section from the Green Line north to Lake Kinneret. The division is not just a political one: the river of the rift runs straight and sluggish through a sandy wilderness, while the river of the valley twists and plays between the hills and pastures of an inhabited land.

The homes here, like those at Beit HaArava, were built precisely because the land was harsh and challenging. Naama, the farming village where Hasson's house stands, is one of a string of Israeli settlements that lies along Route 90 as it runs along the Jordan River and up the West Bank's eastern spine. On official maps and road signs it is called Na'omi. Naama was the name of a settlement in the southern Sinai Peninsula. It was evacuated in the early 1980s as part of the peace treaty with Egypt, just as the first prefab temporary homes of this village were trucked in and placed on cinder blocks.

The settlers wanted to commemorate the lost village of the sands of Sharm el-Sheikh, but Israeli officialdom wanted a different name. It's not good diplomacy to name villages after locations ceded to neighbors.

"They were afraid the old name would give us the evil eye. They suggested Na'omi. We didn't agree. So all the signs say Na'omi, but everyone calls us Naama." Hasson seats me in an overstuffed armchair in her airy living room and serves me lemonade.

"I'm a city girl. My husband grew up in a moshav, a semi-cooperative farming village like this one. After we were married, we lived at Kibbutz Amiad in the Galilee. We moved into the prefabs here in 1982 with a six-week-old baby. The moshav movement was establishing new settlements here. It was a trend then, among people our age, to start a settlement from scratch. The land was parched. We lived for three years without running water. The Lebanon war broke out in June, and the men were called up and the women remained here alone with the babies. It was a very difficult time." She says all this matter-of-factly, and with pride.

Hasson and her fellow settlers in the Jordan rift are nervous and torn. As Israel decides to move out of the Gaza Strip and to evacuate settlements there and in the northern West Bank, she identifies with neither side. Instead, she and her neighbors are relics of an ideological alignment that no longer fits into Israeli politics. The great majority of the Israeli settlers in the Gaza Strip and West Bank are devout Jews. They believe that settling all parts of the biblical Promised Land is a religious imperative and a necessary step toward the arrival of the Messiah. A minority are nonreligious nationalists who believe the Israeli state must include all of the historical land of the Jews. Hasson comes from a different place entirely—from the left, as did the pioneers of Beit HaArava.

"Zionism was nothing to be ashamed of when we came here to settle this region," she told me. It was important for her that I know that Naama was not built over an Arab village. "This was a wasteland. Back then it was called the Land of the Chases, because terrorists infiltrated over the border and the army hunted them down. This spot was a Jordanian army camp. When we first came here, the ground was strewn with berets, bullets, and canteens."

At first, relations with the local Arabs were excellent, she said. "Until the first Intifada began at the end of 1987, we did our grocery shopping in Jericho. I'd go at four-thirty a.m. to Ali's bakery to buy pita bread. We'd take the kids to eat hummus and *ful* at Nabil's falafel stand." Tensions climaxed during the second, most recent, Intifada. "In August 2001, I was driving on the road to Jericho when thirty-six bullets pierced my car. Friends called from Jericho to apologize." But good relations were over. In the same month a kindergarten teacher and driver from the settlements here were killed in a similar ambush. Now the settlers are virtually cut off from Jericho, and Jericho from them.

Naama and the other settlements here were established to secure the border. If the border moves, their raison d'être will vanish. Hasson did not support the Oslo accords with the Palestinians because she doesn't trust the Palestinian leadership. She believes the Palestinians still seek to destroy Israel. She's unsure about the unilateral disengagement plan of Ariel Sharon. The Jordan rift was put on the negotiating table by two earlier prime ministers, Yitzhak Rabin and Ehud Barak. After some initial hedging, both leaders accepted that the rift valley would be part of the future Palestinian state. They concluded that its value as a border was overridden by the opportunity for peace. Hasson doesn't accept that. She'd be willing to give up a good part of the highlands to the west, with their large Arab population. But in her view, Israel needs the line of the Jordan as a security belt against invasion from the east. The rift valley is, in her eyes, instrumental, not eschatological. Therefore, unlike the messianic settlers, she accepts that her government may reach a decision she doesn't like and that she will have to abide by it.

Hasson is very bitter about the Arabs. It's the bitterness of betrayal, which can be felt only by someone who once liked her enemy. "They cause damage. They set fire to fields, destroy crops."

She doesn't rule out the solution proposed by Gandi—transferring the Arabs across the border. Given that she's willing to leave her home, reluctantly, in the framework of a peace treaty, she sees no reason why Arabs shouldn't be required to make the same sacrifice. "The most important thing is to establish a border." The border might be here or there, she says, but the most important thing is for it to be.

From Naama I drive north. The road plunges down and up through bone-dry Wadi Uja, and then through the Arab village of the same name. I take a westward turn off Route 90 on a road that climbs an ancient mountain pass, ascending past strata that mark off the wax-ing and waning of a prehistoric sea. Were it a river (which it some-times is), its waters would plummet from the highlands with hideous force, to be dispersed over the plain below before draining slowly into the Jordan. It ends at the Alon road, which runs south to north along a step in the highlands.

It's a lonely trail in a rocky landscape. Grasses and low-lying trees, greened already by the first tentative rains of winter, make the boulders gentler on the eye than the rift valley below, but I view it mostly uninterrupted by other human beings. My destination is Gitit, an Israeli farming village perched on a small promontory above the rift. The access road leads to a cluster of yellow and white plaster buildings. I take, as instructed, a left, and follow the road around to the Hamo home.

The house is one of several in a crowded cul-de-sac. Children in jeans and T-shirts are tossing a ball around with a dog. The houses look homey but weathered, warm but tired. Eli Hamo emerges in a sleeveless blue undershirt and work shorts, and invites me in. The in-terior is smaller and more cluttered than the Hasson home in Naama. We sit at the kitchen table. He serves me tea with mint leaves tracing slow circles in the glass. In our preliminary introductions, we find that we have something in common. Eli grew up in Hatzor HaGelilit, a town in the upper rift, north of Lake Kinneret, where I lived in 1979 as a young community services volunteer. His family, like a majority of the Hatzor residents, came to Israel from North Africa. But Eli left the town before I arrived. Hatzor was racked with unemployment and poverty. Parents who had ambitions for their children sent them away as soon as they could. Eli's sent him to be a boarding student at a kibbutz school. He enlisted in the Nachal corps, whose soldiers do combat duty and also help establish new Jewish settlements.

"My thinking was to make my life in a new settlement. It was an ideological trend then. My company commander told me that his brother was in a group organizing to establish a new settlement in a beautiful place, a farming community. I'd done farming at the kibbutz. So a week after our wedding, in 1976, my wife and I arrived here with the founding group." The founding families belonged to Beitar, a youth movement associated with the right-wing nationalist Likud party, which would come to power for the first time a year later. They believed in Jewish sovereignty over all the ancient Land of Israel, but they were not religious. They observed many traditions, but they drove and labored on the Sabbath and gave their children secular educations. "Our heads weren't full of theory. We just wanted to work the land."

Liora Hasson told her similar story with fierce pride. Eli Hamo seems troubled. He fidgets, breaks his sentences in the middle to stare out the window, plays with the spoon in his glass of tea. He has the air of a man who is wondering how he'll pay next month's electricity bill.

"Conditions were tough. There was no paved road, and we didn't have enough vehicles of our own. We were dependent on what little public transport there was. The only way to get to Tel Aviv was to go down to the valley, catch the bus to Jerusalem, and then from there another bus to the center of the country."

The young farmers divided up the land between them. There were legal problems—the land was claimed by the Palestinian Arabs of a village farther up in the highlands. The case went slowly through the courts before a judgment favored the Israeli settlers. The Arab villagers are now the Jewish farmers' farmhands. "They have plenty of their own land left," Hamo says.

There were also social tensions among the settler families. The work was difficult. Like all small shareholders, they suffered from unexpected fluctuations in market prices and the vagaries of the rains. Many left.

"In the end, only seven of the original families remained. They were those who 'succeeded,'" Hamo says, making the quotation marks with a twitch of his head. "I'm considered one of those who succeeded at agriculture. We stayed here because we had nowhere else to go."

Israeli agriculture faced major crises in the 1980s as hyperinflation, fiscal collapse, government rationalization of agricultural subsidies, and growing competition from other countries made farming less and less profitable.

"Because of the agricultural crisis, people began looking for work in the center of the country. They commuted from their homes here. But the Intifada made the road dangerous. Palestinians threw stones at our cars, set up roadblocks. Some of our people were hurt, even killed. Four or five families had to leave, and new ones stopped coming. We couldn't maintain our municipal services."

Israel had changed. The "trend" of settlement by young secular families that had brought the Hamos to Gitit had ended. By the close of the 1980s, practically the only people establishing settlements were religious Israelis, motivated by messianic theology. That seemed to be the only pool of new settlers, and without new settlers the village would die.

"A lot of people were afraid they'd take over the village and make it religious." If they turned into a majority, would they ban cars from the streets on the Sabbath? Would they make faces when girls wore shorts and sleeveless shirts? Wouldn't they insist on tightening up the easygoing habits of the village synagogue, used until now mostly for celebrations and high holidays?

The first imposition was the establishment of a yeshiva, a seminary for advanced religious studies. Then, a year and a half before my visit, a dozen and a half young families filled the vacant houses.

"They came here with ideological motives, to bolster the settlement. Some of them work in the center of the country. Others are looking for work locally." Hamo rises to clear away the tea glasses. He rinses them absentmindedly, clumsily, with his head turned toward me. "We advise them not to try farming.

"We have our doubts. Because we don't know where it's leading. Today it's fantastic. But according to the master plan, 250 families are supposed to move in." He shakes his head. "When we came, we had our ideology, but we didn't make such a big deal about it."

Soon after Gitit, the Alon road descends, gradually, with interruptions, toward a pass that cuts through the mountains, connecting the rift valley with Nablus. The fault system here is complex, the landscape a mosaic in flux. The sun has dipped over the western ridge, and I drive in mountain shadow.

Asher Lichtman is security officer for the Jewish village of Hamra. Hamra's geography is challenging, to say the least. The natural road from Nablus, down the valley, has made the village vulnerable to vandalism and terror attacks. The valley also acts as a wind tunnel; gales flog the settlement incessantly. Its farmers have higher overhead because the flowers they cultivate must grow in reinforced hothouses that cost twice as much as the standard screen-form ones.

We sit at a picnic table next to the guard post at Hamra's gate. The wind is moderate now but chilly. It flips the pages of my notebook as I write. Lichtman's body is broad and muscular, though there's some excess weight around the belly that presumably wasn't there when he was younger. His mind seems to be made up of drawers. He opens them one by one and supplies information concisely.

"Labor is a problem. We have workers from Thailand, but the government is cutting back the quotas on foreign workers. We have local Palestinians, too, mostly from the village of Tamoun on the other side of the mountain. We give them a livelihood. Maybe it's not a great salary, but at least it comes in every month.

"Not all the Palestinians who work for us do damage. But there are cases of Palestinians attacking their employers throughout the Jordan rift. We try to treat them with respect. They're human beings just like us. But we'd rather have more Thais. The Thais are excellent workers. They work at full capacity the entire day. The Arabs work in spurts. We know they collect information about Hamra and pass it on to the terrorists.

"We've had two serious incidents. On February 6, 2002, a mother and daughter and soldier were killed when terrorists made their way into the village and broke into their house. On March 6, 2003, two soldiers were seriously wounded in a similar incident. The people here dealt with it, and the terrorists were dead within five minutes. But now the families here are tense. We've had no new families move here since those incidents. There's almost no social life. People don't go out after dark.

"The land here is constantly shifting. Walls crack and floors buckle. The housing ministry has a plan to put down deeper foundations."

The first stars appear as I rise from the picnic table.

"You know the way?" he asks. "It's pretty straightforward. If you make the wrong turn and head toward Nablus, they'll stop you at the army roadblock."

Despite Lichtman's reassurances, a primal fear overcomes me when I turn out of Hamra and onto the deserted road. I am disconcerted when the road turns west, toward the Palestinian city. Soon after, I reach the army roadblock off to the left, where a bored soldier sips coffee and slouches in the penumbra of a concrete barrier. He instructs me to turn east.

Beyond that, the darkness is nearly total. There are no street-lights, except when I pass an Israeli settlement, and those are few and far between. Terrorists could easily lay an ambush behind the boulders along the road. A motorcycle passes me, doing eighty miles an hour on the two-lane road. At the gate to Ro'i, a car waits. A woman is behind the wheel, with two children in the backseat. As I pass, she pulls onto the road behind me. Presumably she has been waiting a long time for a car so as not to drive alone. Another car joins the convoy, then a motorcycle. Then comes a dilapidated vehicle that looks Palestinian, although I cannot see whether its license plate is Israeli yellow or Palestinian blue. None of these people behind me know that this is the first time I have ever driven on this road and that I do not know the way. The trip takes much longer than I expected, but finally Route 90 appears before me. I take a turn to the north and soon cross the Green Line, leaving the West Bank and entering the State of Israel.

The brightest lights in the night are the towns along the eastern side of the Jordan. Much larger than the Israeli kibbutzim and moshavim on this side, they sit on what is practically Jordan's only arable land. On this side of the river, I see the small clump of lights that is my destination: Kibbutz Tirat Tzvi. It lies due east, but I cannot take a direct route. I must go north, past where Route 90 zigzags through a cluster of moshavim, then turn east on a road that curves back to the south. Between 1948 and 1967, Tirat Tzvi was surrounded by Jordan on three sides, located in an enclave that stuck down from the body of northern Israel, along the Jordan River.

Tirat Tzvi is distinctive for a number of reasons. Founded in 1937, it was the first religious kibbutz—the first successful attempt to establish a settlement that combined Orthodox Judaism with the egalitarian socialism of secular labor Zionism. Secularists domi-nated the Zionist movement at the time. They rebelled against their parents' generation, rejecting the authority and constraints of Jewish religious belief and practice, believing that it bound Jews to the backwardness, insularity, and passivity of an exile that was becoming ever more untenable and perilous. They were a comfortable majority among the Zionists in Palestine, setting the tone and public agenda. So the religious socialist Zionist pioneers from Germany and Eastern Europe who erected their tents on this patch of failed farmland on the Jordan were doubly rebellious. In what they called their "Holy Rebellion"—against their parents and the yeshivot where many of them had been educated—they rejected the docility and ossification into which Orthodoxy had sunk. Instead of rejecting religion wholesale, however, they remained loyal to Jewish practice and tradition. In so doing, they rebelled against the dominant stream within Zionism. "We advocate a religious socialism based on the one hand on a recognition of the need for a revolutionary repair of society and on a historical view, and on the other hand on our religious concept and on our in-depth study of the Torah and the social laws it contains," wrote Moshe Una, one of the movement's thinkers, in 1965.

Aside from its special place in Israeli history, Tirat Tzvi is distinguished for some more prosaic reasons. It holds the high temperature record for the entire continent of Asia, a brutal 129 degrees Fahrenheit measured, in the shade, on June 21, 1942, by the commune's small weather station. And it was twice my own home.

I lived here first for two years beginning in May 1982. These were the years of my military service in the Israel Defense Forces. I had no family in Israel at the time and knew myself well enough to realize that if during basic training I had to come home on free weekends to an empty apartment, I'd probably jump out the window. So I gratefully accepted the kibbutz's generous offer of room and board for the duration in exchange for working in the fields for a few months before enlistment. I landed one of the commune's quirkier jobs—responsibility for sex traps against cotton moth pests. The traps, set on poles and adjusted to the height of the plants, were armed with strips of cardboard dipped in female moth pheromones. Male moths made beelines for the salacious odor and died ecstatic, if frustrated, deaths in plastic bags hung underneath. Working largely on my own—which I preferred—I'd tractor from field to field, changing worn-out strips with fresh ones and collecting bags of dead moths. Thanks to my efforts, the bulk of the female moths in Tirat Tzvi's fields remained childless.

I left Tirat Tzvi and returned to Jerusalem after my military service. But I'd made some good friends at the kibbutz, in particular Ronny and Nomi Elon, a couple my age but far more advanced in their family life than I was. They had three children when I met them and eventually had nine. I didn't get married until after I left the kibbutz, and my wife and I never caught up. Every time I visited there with my own family, they'd suggest we come for a year to see whether we might want to live there permanently. Ilana and I liked the city, but the kibbutz certainly had its attractions. There were beautiful lawns and gardens, children running free, and what seemed like a quiet, pastoral life. Finally, in 1990, we decided to give it a try.

Perhaps because its founders had to straddle two ideologies, the small religious kibbutz movement tended to be less doctrinaire and more practical than its larger and secular cousins. During my first period at Tirat Tzvi, in the early 1980s, the kibbutz movement as a whole was in crisis. The crisis had many causes—the capitalist turn in the Israeli economy; the Labor movement's loss of political power, in 1977, to the nationalist-bourgeois opposition; the subsequent decline in government subsidies; and the growing individualist ethos and Americanization of Israeli society. But the proximate cause was that the secular kibbutz movements made a set of foolish investments during the inflationary bubble of the late 1970s. They lost their shorts when the bubble burst. The kibbutzim discovered that they had no way to support their pensioners, the founding generation, who were by then well into old age. Tirat Tzvi was hurting, too, but its senior citizens were secure. Shimshon Caspi, the kibbutz treasurer, explained to me one night as we were on guard duty together that the religious kibbutzim had years ago had the foresight to pay into pension funds and savings accounts for the future welfare of their founding generation. The secular kibbutzim had assumed that the younger generation would work to support the older one. When it became clear that a lot of the younger generation were leaving for the cities or for other countries, they thought the state would make up the shortfall.

Among the religious kibbutzim, Tirat Tzvi was one of the less hidebound. When we were there as a family, big changes were already under discussion. More and more people wanted to take jobs outside the kibbutz, and many were angry that the commune placed tight restrictions and time limits on independent careers. As agriculture became less profitable, hardworking, ambitious members with professions became increasingly aware that they were working overtime to support some laid-back colleagues. Hardly anyone at the kibbutz wanted to work in the sausage factory that was the commune's major earner, so workers had to be brought in from outside.

I'd heard that the kibbutz had made some changes in the fourteen years since we left. But I'd been out of touch, and I wasn't prepared for the cash register in the dining hall when I met Ronny there for breakfast on Monday morning. A middle-aged man dressed in the standard blue work shirt and shorts pressed buttons to record each item we'd taken and deduct the cost from Ronny's account. I could see that people were taking smaller portions than I remembered—one egg instead of three, two slices of bread instead of eight. I told him that Ilana and I had once been shocked by how much food got thrown away. "Not anymore," he says.

I call one of Ronny's two cell phones when I arrive at the entrance to the kibbutz the previous night.

"Great timing." He laughs. "I'm out plowing, but I'm almost done. You can come pick me up in the field and take me home. Do you remember where the reservoir is? Head out there and I'll see you."

Instead of turning right into the residential section of the kibbutz, I turn left and drive to the fields. There I follow a one-lane strip of asphalt as it circumvents the kibbutz, past fishponds lying still and dark in the night, past a fallow field and a small cemetery. My cell phone beeps; it's a message from a Jordanian provider informing me that I am now on their network. If I call Ronny now, it will be at international rates. That's how close to the border I am.

The asphalt turns away from the kibbutz fence and continues south between furrowed fields where low-lying greens sprout. When I reach a clearing near the reservoir, I hear the drone of the plow. Ronny telephones me to say, "I see you, I'm coming your way," and soon I spot a pair of headlights coming in from the west. The lights stop at a safe distance, then go out. The engine falls silent, and Ronny climbs out of the crow's nest and walks over to my car.

"They needed people to plow at night, and I volunteered. I enjoy it," he explains. "And I like to do some work on the farm."

When I was here as a soldier, Ronny worked with me in the field crops division, growing cotton in the summer and wheat in the winter. From the start, he and Nomi amazed me by their ability to work long hours on their feet, tend to their growing family, serve on kibbutz committees, maintain large social networks both on and off the kibbutz, take classes, and even read some of the books that Ronny was constantly adding to their huge library.

In response to demand from the membership and the need to make farm operations more efficient in an increasingly competitive economy, the kibbutz has changed its employment policy and management structure. The different agricultural divisions—field crops, fishponds, turkey run, date groves, and meat factory—now operate as independent economic entities that must show profit. Some of them are managed by outside professionals rather than by members. On the one hand, members of the commune are no longer guaranteed employment in one of the kibbutz operations—in theory, at least, they must compete for job openings along with outsiders. On the other hand, members are now free to pursue their own careers. Working outside the kibbutz is no longer a special privilege that must be approved by the entire membership. All can work where and at what they please, so long as they earn at least as much as they are required to contribute to the general kitty. Ronny juggles several outside jobs—he works at the Religious Kibbutz Movement yeshiva up on Mount Gilboa, at a pre-army work-study seminar in the Jordan rift valley, and at a few other places.

"We're still a commune," he tells me as we pull up by his red-roofed, white-plastered, earth-hugging duplex not far from the perimeter fence. "Whatever you earn goes into the general fund, and you get a budget based on your family size. And the kibbutz still supports a lot of community and charitable projects and funds higher and continuing education for members."

"So it's sort of living in a close-knit, community-conscious small town with very high taxes," I say.

When I came with my family for a year, I envisioned a less hectic pace than Ronny and Nomi maintained. I thought I'd have time to sit out on my lawn and read, time to chat with the elderly founders of the kibbutz. But work in the turkey run turned out to demand a lot of time, not always at conventional hours. The rest was taken up by two small children and a baby born while we were there. So when I planned my trip up the rift, I blocked out a day for Tirat Tzvi to do what I hadn't done years before.

Early Monday morning I report to the kibbutz archive, housed in one of the long, barracks-like structures that were among the first built by the settlers. The archivist is Yizraela Caspi, wife of Shimshon, the former kibbutz treasurer. She is also a native, one of the first children born on the kibbutz. In the early 1970s, Yizraela became the first woman to hold its highest office, internal secretary, and has served in a number of other leading positions on and off the kibbutz as well. She's got short black hair and the voice of a warmhearted schoolteacher who nevertheless knows how to make her pupils toe the line.

I begin reading through the material she takes off the shelves for me. I learn that the land on which Tirat Tzvi was built was bought by the Zionist movement from three Arab landowners, one of whom was Musa 'Alami, an important Palestinian nationalist leader in Jerusalem. He called on his people to battle Zionism to the death and excoriated Arabs who sold land to the Jews.

Geologically, the Beit She'an valley is a basin. Its floor subsided as the rift opened and as the trans-Jordan landmass moved northward. As a result, it was not a desert, but it became a waste of water. Springs, both salt and fresh, well up throughout the valley, but left to their own devices they create malarial marshes. In ancient times the metropolis of Beit She'an—Scythopolis to the Romans—prospered at an important junction in the road system, a short way north of Tirat Tzvi. Its marketplace was cooled by large fountains, fed by springs that were channeled into the city. When the Muslims conquered Palestine, the city continued to prosper, but it declined after a serious earthquake in 749. When the city disappeared, so did control of the water supply. The valley reverted to marsh and wilderness on the edge of civilization. Nomadic raiders ravaged the remaining population and eventually came to control much of the valley's lands.

When the Zionists began buying up land on which to settle Jewish farmers, Arab speculators began purchasing land as well. The Bedouin sheikhs of the Beit She'an valley were happy to sell, sometimes in the hopes of bettering the lot of their tribes but no less often tempted by the opportunity to make a personal profit. The wealthy Arabs who bought the fields near the river hired locals to cultivate them, and built a small, walled enclosure with a farmhouse. But malaria and Bedouin raids made the venture unprofitable, and the owners were more than willing to sell the land off to the crazy Zionists.

The religious Zionist movement had been waiting for an opportunity to establish its first kibbutz. In the wake of the Arab rebellion of 1936–1939, the British announced severe restrictions on Jewish immigration to Palestine. The Zionists decided to respond with the "Stockade and Tower" operation, in which they established new settlements throughout the country. Three groups of aspiring religious Zionist communal farmers came together to establish Tirat Tzvi.

"I came four years after the kibbutz was founded." Yitzhak Ariel sits by his bed in a wheelchair that supports his back and head. "There were two shacks and some tents. I looked on the founders as if they were giants. I came from Tel Aviv."

He wears a green shirt with blue polka dots, blue pants, and the standard kibbutz sandals. Despite his illness, his broad shoulders and sinewy neck still bear testimony to the half century he spent working in the fishponds, hauling huge nets of carp out of turbid waters. I remember him behind a similar but less sophisticated chair, wheeling his wife into the dining hall on Friday nights after the Sabbath prayers. She lived for three decades with Parkinson's disease. I don't know if that's the disease that has attacked him; I am embarrassed to ask. I came to visit him in the kibbutz's assisted living facility, right next to the old two-story building where I lived when I was a soldier. Its founders and old-timers aging, the kibbutz built this small, homey building after I left.

"I found people with values. They dug irrigation canals with picks and shovels between springs and fields in impossible conditions. I came from a home with nice silverware and furniture. I almost collapsed from the hard labor. I remember picking cucumbers by hand for hours and hours. But my luck is that God made me the kind of person who is always satisfied with whatever he has." His smile, full of good humor, seems to bear a silent message: there are worse things in the world than to be paralyzed.

He's not at all worried about the cash register in the dining hall. He sees the changes not as a loss of the communal ideal but as its necessary adjustment to modern realities. "To this day Tirat Tzvi succeeds at being a kibbutz. Look at how I'm cared for here. Making people pay for the electricity they use is completely in keeping with the communal idea. Before, when electricity was free, people left their air conditioners on all day. Now that it costs them money, they turn them off. There's less waste."

Yitzhak doesn't see himself as one of the founders, but for members of my much younger and lesser generation he, too, was one of the giants. The men and women of his generation worked long hours in infernal heat, were dedicated to their families, and could still find the time and strength to study a page of Talmud in the evening. They were strict in their observance but wore religion lightly, not because it was unimportant but because it was integrally part of every minute of their lives; it did not demand gravity or ostentation.

While they shared a communal religious Zionist creed, they varied in their politics. Most of them believed that the establishment of the Jewish state was the first stage in a process that would lead to the Messianic era. Some were militants, who believed that the end of days was fast approaching. They believed that Israel must expand to include all the territories of the ancient Jewish kingdom. But most, like Yitzhak, were easygoing messianists, who believed that God's kingdom would be established by promoting justice and equality within Israel's territory, not by the expansion of that territory. It's hard to imagine it now, but there was a time in which religious Zionism's political leaders included some of the country's leading doves. Their government ministers voiced outrage when Israeli soldiers killed Palestinian women and children in antiterrorist raids. Yitzhak doesn't want to talk politics with me, but the only time he frowns during our conversation is when he mentions his "crazy son," who is a member of the Knesset for a far-right Greater Israel political party.

On my return to the archive, I encounter someone who remembers the old days quite differently. Shlomo Yorav sits in a small tractor. Hitched up to it is a wagon full of odds and ends that he is hauling from one side of the kibbutz to the other. He has stopped by the archive to ask Yizraela an incoherent question, rendered utterly incomprehensible by the engine's growls. She introduces the two of us and explains my visit. Yorav fixes me with a desperate gaze full of an energy not evident elsewhere in his aging body. Round-faced and thin-haired, he wears a blue work shirt unbuttoned down to the belly. He and Yizraela are almost classmates. He speaks frantically, as though he has been waiting for me many months.

"Someone said, I can't remember who, that some people forget, and some people remember only the good things, but the madmen remember everything. I remember everything. It's a problem to teach you this story because you know where it ends, but where does it begin?

"We children lived in the old plaster farmhouse the Arabs left behind. There was a room for boys and a room for girls. We had the walls of the farmhouse around us, beyond that three fences, beyond that the mountains closing in on us. We were closed in on all sides. And for what? All in the name of a mad ideology. We were the continuing generation, the children who would carry on the dream. We would do what our parents and the movement decided we should do. Everything was collective, like an army camp.

"Once we went to visit Kibbutz Yavneh, a religious kibbutz down in the south. I couldn't believe it—the children walked around freely, no one stopped them, pulled them back. Kids would hop on a wagon, and no one counted them or took account of who was going and where and for how long. They took us to Mount Sodom, that hill of salt next to the Dead Sea. Can you imagine? For us it was like going to Mars. Whoever heard of such a thing, going out and walking up a mountain as if everything was allowed? Once, when I was about six, my parents took me to visit Jerusalem. We stayed with family in the Bukharian neighborhood. I went outside and began walking. I walked and walked and there was no wall or fence to stop me, so I kept walking.

"Jerusalem was to the south, but for us the only way out was north. We only went north. I could never understand why we faced south when we prayed. What was in the south? It was all a mystery."

"So why has he stayed here all these years?" I ask Yizraela later, as she serves me coffee in her small home.

"He stayed." I recall Amotz Zahavi's take on the choice between altruism and egoism. Maybe Yizraela means that it's more pleasant to be an eccentric altruist on a kibbutz than a madman in a city.

Yizraela's memories are different still. She fondly recalls her childhood, growing up in a children's house rather than with her parents. When, between high school and the army, she went to study for a year in Jerusalem at a religious teachers' college, her status as a kibbutz girl earned her awe and respect from students and lecturers, as well as special dispensations: only she, of all the girls, was allowed to wear pants.

What she calls "creeping privatization" began in earnest a decade ago, after a small group of "very strong" families—people who'd been extremely active and had held senior positions—announced they were leaving. They no longer saw any value in a communal lifestyle and wanted to pursue their own careers without constraints.

"It was a great shock, but I decided not to allow it to break me," she says. The kibbutz held a series of assemblies and reached the fiscal decisions whose effects I've been noticing. "We decided people had to pay for their electricity, their food, their purchases in the kibbutz store. Everyone has to bring home their own livelihood now."

Almost everyone is gainfully employed, she maintains. "About three percent of the people here have a problem. They can't find work, and the community supports them. But that's the same problem any society has." Agricultural pursuits—field crops, vegetables, and Yitzhak Ariel's fishponds—which were once the very essence of the labor Zionist ideal, are struggling to show a profit and may be phased out.

"I'm not sure that there will be a kibbutz here in the future. I'm not sure it interests the young people here. They live different lives now. Their leisure culture is different than ours was. The young women are pulling in a more religious direction. People don't wait for the community to organize activities, and people don't show up for the ones we have. Shimshon is an ideologue and thinks it's all wrong. I accept it, even with some empathy. When we stopped eating in the communal dining room, I discovered the pleasures of having a family Sabbath meal at home."

The next morning is Tuesday, the third day of the week, when God established borders between dry land and sea. Oddly, this day of separation is traditionally the favorite day for Jewish weddings, since it's the only day on which God twice "saw that it was good." On this particular Tuesday, the Knesset will vote on Israel's separation from the Gaza Strip and a section of the northern West Bank. Despite a rebellion within his own Likud party, Prime Minister Sharon will win this vote, with the help of the opposition.

I drive north on Route 90. The road is straight, but it rises and falls with the undulations of the underlying earth. It is this section of the rift, from Beit She'an up to Lake Kinneret, that is called the Jordan Valley in Hebrew, or more precisely the Central Jordan Valley. It is a narrow passage between western mountains and the river, and the eastern mountains beyond. Ancient travelers on the great Sea Road from Egypt walked and rode along the rift here, fresh from the inns and theater in Scythopolis and ready for the trip up around the lake, then across the Golan plateau to Damascus.

It looks green and fertile today, as it was in the ancient period. Polybius wrote in the second century B.C.E. that the region between Scythopolis and Lake Kinneret could supply food for an entire army. It continued to flourish during the first century of Muslim rule, but it later began to decline, perhaps because of a combination of earthquakes and politics. Damascus was the first Muslim capital, but when the political center moved to Baghdad, traffic declined. The Crusaders built a fortress, Belvoir, on a peak just west of the road at the center point between Beit She'an and the lake.

The valley hides the scars of the nineteenth century. The Turks were weakened by the Crimean War, which was accompanied by drought. Bedouin tribes, starving in the eastern desert, invaded the valley. In the later part of the century, the sultan tried to revive his fortunes and reinforce his rule by resettling North African Muslims—Moroccans and Algerians fleeing the French invasion—here. The incentive he offered them was exemption from military service. Sudanese laborers were also imported to work in the fields. But agriculture did not prosper. While there was ample water in the Jordan, the river was far lower than the fields, and the Muslim farmers had no mechanism other than their own shoulders and their donkeys' backs to make the water flow uphill.

Route 90 crosses a bridge. The Jordan River flows underneath; here, for a few kilometers up to Lake Kinneret, Route 90 runs to the east of the river. Just south of the bridge, the Yarmouk River flows into the Jordan. The Yarmouk, which now forms the border between Israel, Jordan, and Syria, is much more ancient than the Jordan of the Bible. Before the rift opened, it flowed mighty and broad, straight into the western sea. Route 90 now passes through low-lying tropical meadows. Bananas are the salient crop of the Kinneret littoral, and double trailers stacked high with huge stalks of hard, green fruit pull out of side roads and onto Route 90.

One of those side roads leads to Kibbutz Sha'ar HaGolan. I pass by corrugated metal sheds, tractors, and a water tower—it's as if the kibbutz's first inclination is to show off its own industriousness. But except for the banana trucks, there doesn't seem to be much going on. There's one telltale sign of change, though—in front of a shed is an army of two-wheelers and a sign: "Bicycles for Sale." In the old days of HaShomer Hatza'ir, the most radically Marxist of the kibbutz movements, to which Sha'ar HaGolan belongs, bicycles would have been held communally.

I've come to see this place not because it sought to be a twentieth-century communist paradise but because of the kibbutz's ancient predecessor. Eight thousand years ago a village prospered here. Its inhabitants didn't grow bananas. Judging by the large number of clay and stone figurines unearthed here by the archaeologist Yosef Garfinkel and his staff, the major industry could have been the production of idols. Garfinkel's excavation owes much to the decline of the kibbutz. From objects found on the surface, archaeologists knew that the site deserved excavation, but the area was for a long time covered by olive groves and fishponds. When those were abandoned as economically unviable in the early 1990s, the scientists were able to move in. Garfinkel completed a decade and a half of excavations last summer and covered up the site, so there's nothing for me to see in the field. But the kibbutz runs a little museum where visitors can view some of the finds. It's a one-story white building that lies just beyond the kibbutz houses and lawns, a few steps above the adjacent field.

The website says it's open every morning from nine to twelve, but the door is locked when I arrive. A scrap of paper taped above the handle tells me to call Rafi and offers a cell phone number. Rafi comes straight over. He turns out to be a thickset guy of about sixty, dressed in a regulation blue work shirt and shorts with holes in them. He's a bit surprised to see me alone; apparently he thought he was coming to open the place for a busload of tourists. He stares at the kipah on my head. After fussing around a bit to take my money, sell me a book, and write out a receipt, he unlocks the door to the exhibits.

The room is small but well-organized. The cases are labeled and offer clear, concise explanations in Hebrew and English. The room is set up so that a clockwise circuit takes you through what the archaeologists have learned from the site. The Neolithic residents of the prehistoric village belonged to what's called (after the river) the Yarmukian culture. The Yarmukians postdate the Natufians, whose worked standing stones predate the unworked ones that so fascinate Uzi Avner. The exhibits contain sections about the houses the Yarmukians lived in, the utensils they used for cooking, dyeing, and other functions, burial customs, trade, and figurines.

But Rafi insists on taking me counterclockwise through the exhibits, from back to front. And instead of letting me read the labels, he throws me a half sentence about each case, then pulls me on to the next. When I dawdle, he gets impatient.

As I try to get a good look at the array of figurines, Rafi tells me about his eldest son. "He's become one of you people, but the most extreme kind," he said. "A Bratislaver Hassid. He studies at a yeshiva in Safed. He thinks he can get the Messiah to come."

I ask a question about some pebbles with lines engraved on them, but Rafi has lost interest.

"Do you study kabala?"

"No," I say. "It doesn't interest me that much."

"What language do you study kabala in?" Rafi wonders out loud. "That book, the Zohar, what language is it in? My son said it was written by a rabbi named Shimon bar Yochai, who lived in a cave."

"It's in Aramaic," I say. "Actually, an artificial Aramaic with medieval influences, since it was written in the Middle Ages."

"He thought it would solve all his problems, I guess," Rafi ponders. "He wouldn't need to work. Just study."

"Can the archaeologists know if these figures were worshiped? How can you tell it's an idol and not lawn sculpture?"

Some of the figurines have a clear, if very stylized, human form. A rounded stone displays, despite the erosion and breakage of millennia, folded arms, a navel, two breasts. A long, slender one has two legs, a phallus, and testes. Others are much more enigmatic— the label says this is a woman's behind, but it could just as well be a front or a lump of stone.

"He says it's like they gave him a new memory. Like his brain is a computer disk and they replaced it. He's written a book. I'm trying to help him get it published. Maybe you know someone who wants to publish a book on the Messianic age? I think that's what it's about. Believe me, I can't understand a word of it. But it's important for him, and he's my son. I was born here when the kibbutz was young. My parents thought their descendants would live here for generations, new Jews who didn't need God, only labor and the land. My younger son is studying physics at the Technion. Their mother and I are divorced. I was wounded in the Six Day War. I can't work long days outside. So I sit here in the museum."

PART III

FLOATING IN THE AIR

Aland becomes mythical in three ways. When it is distant in place or time, we imagine how it would be if we were there. If it is the land we live in, we attach to the places around us our stories about who we are, where we came from, and where we are going. And if we lived there once, we inscribe in it stories of the tribulations and exhilarations of our younger, half-forgotten selves.

Soon after I arrived in Hatzor as a young volunteer at the beginning of 1979, I walked to Tuba. Most of the people I worked with in the Jewish town were confounded by my interest in this Bedouin village on a hill three miles east of Route 90, not far from the edge of the fault line where the Jordan flows in a rocky, narrow channel down to Lake Kinneret. Some warned me sternly that I was taking my life into my hands. But I was curious about the people who lived on the ridge, simply because they were so different from me. Twenty-two years old, I didn't really believe that anything awful could happen. Rain began falling soon after I crossed the highway, and by the time I got to the turnoff, I was soaking wet.

The first house lay some way down the narrow asphalt, which was more a path than a road. It was a shack of corrugated aluminum set on a concrete platform. A five-year-old girl in a stained dress gazed at me, wide-eyed, from the doorway. A few similar shacks followed. The asphalt was cobbled with cow pies, and the air smelled of wet sheep. Some more substantial plaster-and-concrete houses followed the shacks, and when I reached the village's central junction, there was a cluster of large, new residences. They belonged, I would later learn, to the mayor, the elementary school principal, and the head of the workers' council. Across the road, on the ground floor of a half-finished concrete shell, a door stood open and revealed the counter of a sparsely stocked grocery. A surly man with a spotty mustache and a creased face stood behind the counter. He was handing two packs of cigarettes to a slightly built teenager.

I must have been a bizarre sight, dripping rain in the doorway, wearing blue jeans and running shoes, but I was oblivious to that.

"I'm Haim. I'm from America," I said in my wobbly Hebrew. The storekeeper stared at me and, as if certain that I could not be real, turned to count the coins in his cash drawer. As if it were the most natural thing in the world, however, the youth held out his hand and in accented but competent English said: "I'm Naser. Please come to my house to have coffee."

Naser was in high school then, meaning he'd been born in the early 1960s, the heyday of the pan-Arab enthusiasm led by Egypt's president Gamal Abdel Nasser. Abdel Nasser means "Servant of the Victorious One," and the name Nasser or Naser was synonymous with Arab pride. The Egyptian leader promised to turn his backward country into a modern one, to restore the glory of the Arab nation, and, of course, to destroy Israel. Naser is a common enough name. The Heib Bedouin have a history of loyalty to the Jewish state, and some of the village's young men volunteer for army service. Maybe my friend's parents didn't have the Egyptian leader in mind when they chose a name for their son. But when a name, even a common one, is attached to a person who makes history, the name remains the same and its resonance changes.

Naser led me farther into the village. The rain had subsided, and along the way he picked up some friends who were starting to emerge from their shacks—Hussein, a muscular black kid; Jihad, a downy-faced boy in tight jeans; and some others whose names I don't remember.

Naser's house had two living rooms—one with a couch and armchairs, in Western style, and one furnished more simply with carpets and mattresses along the walls. That Naser ushered me into the latter was a customary sign (though I didn't realize it then) that he was treating me as a peer rather than as a formal guest. The message came across quite clearly even without that knowledge. Naser's younger brother joined us, bearing a tray holding glasses of Turkish coffee. His sisters and mother remained invisible except for an occasional head peeping in to see whether we needed anything more.

The boys were all around sixteen years old. Some, like Naser, went to a high school in an upper Galilee village beyond Safed. Others were already out of school and working as tractor drivers for Kfar HaNasi, the kibbutz on the ridge to the east overlooking the Jordan gorge, in the restaurants down at the Rosh Pina junction, or as seasonal laborers in the region. Naser was clearly the sharpest member of the group, well-read and thirsty for knowledge. But he was also a cynic, even a nihilist. He was disgusted with the rigid tradition of village life and what he believed were the outmoded and senseless demands of the Islam he had been brought up with. He disliked politics intensely and thought the Israel-Arab conflict was little more than brutish bloodshed. So he wanted to get away, to some other country, as soon as he possibly could. He vowed that he'd never marry and bring children into such a bleak world.

His worldview was, in short, almost diametrically opposed to my own, but we became good friends, perhaps precisely because we were intrigued by the contrast. One fine spring day a few months later, when I was on another visit at his house, I asked him what lay on the other side of the ridge on which Tuba and Kfar HaNasi stood.

"Down there?" His face indicated that it held little interest for him. "Just the Jordan."

"Why don't we take a walk down the hill?"

He shrugged his shoulders. "It's steep and uncomfortable, and there's nothing there to see."

When I walked through the rain to Tuba, it didn't seem remarkable. But that walk has become a story I have told myself and others many times since. I now realize that it says a great deal about who I was when I was a young man. My twenty-two-year-old self is very distant from me now; when I see photographs of myself at that age, I am surprised. The face and figure don't seem connected to me. But when I recount my walk to Tuba, I know that who I was and who I am now are one and the same.

The drive from Sha'ar HaGolan to Lake Kinneret takes a few minutes on Route 90. As I drive, I hear the broadcast of the Knesset's vote on the disengagement plan. It passes. The road turns west to skirt the lake. I cross the Jordan again, at the point where a dam holds Kinneret's water in. A short way beyond that, on a small mound between road and water, I see Rachel's grave.

The best-known tomb of Rachel lies on the outskirts of Bethlehem. Today the site is a point of contention between Israel, which wants to keep the holy place under its control, and the Palestinians, who see it as part of Bethlehem, a Palestinian city. The Rachel commemorated there is Rachel the Matriarch, the younger and most-loved wife of Jacob. The tradition that the site is Rachel's grave goes back many centuries; the book of Genesis says Jacob buried her "on the way to Efrat, which is Bethlehem." But there is another Bethlehem in the Galilee, not far from Sepphoris, a city that was the Galilee's capital before Tiberias was built. Here, in this drowned diamond graben of the rift valley, there is another grave, belonging to a different Rachel.

In a quarter century of traveling north and south on Route 90, I never before stopped at the cemetery between the road and the sea. Today I pull in to the tiny bulge in the road's shoulder that serves as a makeshift parking lot. The gate is open. A bus follows and blocks my car.

A grove of date trees stands to my left, just beyond the cemetery fence. Its fronds, thirty meters up, sway gently in the breeze blowing off the water. The trees, too, are immigrants. They or their progenitors were brought here some decades ago from the shores of Sha'at al-Arab by a man who lies buried inside.

Before me is Lake Kinneret, the tiny sea of the Galilee, its surface clothed with a layer of mist that separates it from the sky. I cannot see its northern shore, but I can vaguely make out the clutter of Tiberias halfway up the western coast. The shore here is rocky, with willows growing up to its edge, shading the headstones and the lakeside. They seem to absorb all sound; despite the highway just a few paces away, all I hear is the lapping of waters, the whisper of leaves, and the rustle of gravel under my feet.

I walk up a few steps to the mound where the graves lie. The early Jewish settlers who, on the fields across the road, invented the concept of the kibbutz, chose this hill to inter their dead because it is a peaceful spot with a view of the lake.

The bus has disgorged a load of high school boys. They jostle through the gate as I head down a row of headstones halfway up the tell's gentle slope. Toward the end of the row, just a few feet from the shore, I find three headstones that tell a story. Berl Katznelson, one of the founders of the labor movement in Israel and the first editor of its now defunct newspaper, *Davar*, lies buried next to his wife, Leah Meron. To the right, close enough to be part of the group but separate nonetheless, is a smaller headstone, inscribed "Sarah Shmukler." Katznelson and the two women immigrated to Palestine at the beginning of the twentieth century and were a close, yet chaste, threesome. They understood tacitly that the romance was between Berl and Leah, even if they often lived far apart. But once when Berl went to visit Sarah at Yesod HaMa'alah, a settlement next to the Hula swamp, they fell in love. A short while later, Sarah came down with malaria and died in the hospital in Safed. She was buried here, and Berl and Leah were married. Both were now wounded—Berl mourning his true love, and Leah stung by the betrayal of her two soul friends—and the marriage was not a happy one. The headstones seem to reflect this; their gaze is permanently fixed on the best friend, lover, and rival, who lies apart, far enough away to be alone but near enough to be forever bound up in a web of jealousy.

The boys fan out through the cemetery despite their teachers' attempts to herd them behind their guide. Angular and aging, he wears a broad-brimmed Australian canvas hat and old-fashioned sandals. The three teachers, earnest young men with short black beards, call out "Stay with Yossi, he'll show us around." The boys are most interested in finding the newest grave on the mound, that of the songwriter Naomi Shemer. She was born on Kibbutz Kinneret and spent her childhood there. Later, she became the lyricist and composer of dozens of songs about the Land of Israel, many of them laden with nostalgia for the pastoral pioneer days at Kinneret and similar spots in the north.

The boys who gather at her grave are wearing kipot, and I learn that they are from a high school yeshiva in Hispin, a settlement on the Golan Heights. Shemer was not religious, and she lived most of her life in Tel Aviv. Ostensibly, she did not have much in common with these boys. But in modern Israel, ironically, the subculture that identifies most strongly with Shemer's nostalgia is one that was never part of the world that she sang about. Her songs of the land stir religious boys whose grandparents were shocked and dismayed by the atheistic socialism of Kibbutz Kinneret. The boys are now adding stones to the pile on top of the grave—a traditional sign of respect for the dead.

One boy wanders to the only other grave where the heap of stones is so large. It's the one that I came here to see, and its very shape signifies that it's the focal point of the entire cemetery. It's not the largest marker, but it is set apart. An L-shaped extension of the gravestone serves as a bench where a visitor may sit and contemplate. He may also read; the boy removes a heavy metal lid from a cavity in the bench and draws out a dog-eared, coverless book. I have in my hand a copy of the same book, which I rescued from a box of discards a neighbor was throwing away a couple of years ago. It's bound in white, and the title imitates the poetess's handwriting: *The Songs of Rachel*. Inside is a photograph of Rachel Bluwstein, 20 September 1890 to 16 April 1931. On the overleaf is a facsimile of a poem in her own hand: "Only of myself I know to tell/My world is as narrow as that of an ant." The poem's title is "A Locked Garden." The gravestone says only "Rachel," for so she signed all her poems. Few people know her last name. In Israel, where her poems are part of the high school curriculum and are set to music by popular songwriters, she's universally known as "Rachel the Poetess." Her namesake, entombed near Bethlehem, is "Rachel the Matriarch."

"Can we all be quiet?" asks one of the teachers after he and his two colleagues have managed to gather up the boys and sit them down on the boulders, flagstones, and terraces around Rachel's grave. "Yossi wants to tell us a story."

Yossi settles down onto the bench with his back to the reading boy, places his elbows on his knees, and presses his fingers together. He surveys his audience and begins.

"My story is about a Jewish boy named Joseph. His father was a very serious Hassidic rabbi from a very well-off family. Joseph was an only son. One day when he was a young boy on his way home from school, he was kidnapped by Russian soldiers. As was the practice then, they banished him to a remote village to be brought up by a Christian family, in the hope that he would forget he was a Jew."

But Joseph did not forget, even when he reached military age and was drafted into the army. Yossi reminds the boys that the Russian army's mandatory term of duty for young Jews was twenty-five years. Unlike most Jewish conscripts, Joseph turned out to be adept at the military profession and rose in the ranks, all the while observing Jewish religious precepts as best he could. He participated in the Crimean War with valor, but his refusal to convert to Christianity blocked his advancement as an officer.

Yossi shrugs.

"Joseph was Rachel the Poetess's father. When he completed his twenty-five years of service, he took a wife and began to sire children at a frantic pace. His wife died giving birth to his fourth child. He married again and produced eight more children, in between the long journeys he took as part of his successful gemstone business. Rachel was his tenth or eleventh child. The family had close ties with the czar's court. All the children received a traditional Jewish elementary education and then were sent to secular schools and to college. Rachel and her closest sister, Shoshana, were headed along the same path. Rachel was a talented artist and Shoshana a musician, and Rachel was supposed to study art and philosophy in Italy. But the two sisters decided to take a trip to Palestine before continuing their studies." It was 1909; Rachel was nineteen years old.

"The Jewish community in Palestine at the time, the beginning of the Zionist migration, was tiny but vibrant, and the girls were soon caught up in it. They settled in Rehovot, then a Jewish farming village, where they worked in the citrus groves and turned the boardinghouse they roomed in into a cultural center.

"They hosted a kind of salon every Friday night, but they had a problem. The girls didn't speak Hebrew, and none of the Zionists were willing to talk in Russian, even though most of them knew the language very well. So Rachel and Shoshana went to the village kindergarten and asked the teacher if they could enroll. They sat on the benches with the little tykes, and that's how they acquired the language—from the bottom up. You see that in her poems. The language is simple, down-to-earth, and the images are very sharp, the way a child sees the world."

Yossi tells the boys about Rachel's passionate but ultimately disappointed first love for Nakdimon Alschuler, a handsome Jewish pioneer.

"Everything was fine until one bright day their door opened and in walked A. D. Gordon, who began to say strange things. Such as, one should not work for sustenance but because of its value for the spirit." Gordon was indeed an exceptional man, even in that Jewish immigrant society in Palestine, which harbored more than the usual share of prophets and eccentrics. He wore a Tolstoy-like beard and preached a labor gospel of working the land. Under his influence, Rachel moved to Hatzar Kinneret.

Yossi speaks dramatically. He looks one boy and then another in the eye. He tells them that Rachel's time at Kinneret was the happiest of her life. She worked long hours on the land but was at one with the trees, the fields, and the sea. There was love—by some accounts she had a romance with Berl Katznelson, who lies not far from her with his two women. A young Lubavitcher Hassid turned Hebrew Marxist named Zalman Rubashov would later write that he fell in love with Rachel at first sight when he came to visit Katznelson at Kinneret. Half a century later, after changing his name to Shazar and moderating his revolutionary zeal, Rubashov would become Israel's first minister of education and third president. His discarded name wanders into Arthur Koestler's *Darkness at Noon*, a novel about the corruption of the Marxist myth. Shazar's account is not reflected in Rachel's poetry: many years later, she dedicated "A Locked Garden" to him. It's a poem of unrequited love.

I'd always imagined that she worked Kinneret's fields by day and wrote her lyrics in the evening. But Yossi says the poems were actually written much later, mostly in Tel Aviv. I realize that this should have been no surprise. The landscape we see before us is never as vivid, or as in need of words, as the landscape in our minds.

Having learned what she could at Kinneret, Rachel decided to leave the lake and study agronomy in France. World War I exploded before she completed her studies, and she was unable to return to Palestine. She made her way to Russia, where she taught Jewish war orphans. After living there through the Revolution, she returned to Kinneret and cared for young children. Soon thereafter, she was diagnosed with tuberculosis. The disease was contagious and incurable.

Yossi stiffens. "The mothers said, we will not allow her to work with our children. She must leave. 'At nighttime came the messenger, and on my bed he sat,' she would later write. 'He threatened me with wasted fist/Let out an evil laugh/The poetry comes now to its end/This song your very last.' When she left with her belongings, no one accompanied her or bade her farewell. They were all working in the fields. She wasn't angry with them, and she continued to write to her friends there. But she never visited.

"She went from Kinneret to the hospital in Safed, where she stayed a few months. It was there that she began writing her poems." Later she moved to Tel Aviv, where she wrote verses that appeared weekly in Katznelson's *Davar*. She also wrote occasional essays and criticism. In a review of the play *Six Characters in Search of an Author*, she quotes the playwright, Luigi Pirandello: "Reality, with all its baseless things, large and small, has an immeasurable advantage over art . . . The insubstantiality of life doesn't need to be shown to be true, simply because it is true."

The boys closer to Yossi are listening attentively; the ones farther away are fidgety. Some of the boys in the back are engaged in a deliberate display of insouciance, an attempt to show their buddies that they have no interest in poetry whatsoever—poetry being for artsy-fartsy, intellectual left-wingers rather than for sturdy young men who will be in the army in a couple years' time. In any case, these boys are students at a yeshiva, where they are taught that the Talmud and midrashim are the texts worthy of their attention. They are not forbidden, but neither are they encouraged, to read secular literature. The tomb of Rachel the Matriarch near Bethlehem holds a larger place in their consciousness than the grave of Rachel the Poetess here on the Kinneret's shores. Still, their teachers have brought them here to hear Rachel's story, in Yossi's words.

As I listen to Yossi, I sense that he is seeking a niche in the boys' minds. Am I imagining it, or is he deliberately implying a parallel between the two Rachels? Is he comparing the poetess, lying in her deathbed in Tel Aviv, with the matriarch, whose soul leaves her during childbirth? Is he conflating the poetess's sorrow at her exile from Kinneret with the matriarch's lament for her children, as they pass her tomb en route to exile in Babylonia? Is he bringing Judea to the shores of the Tiberias lake? After all, that Galilean Bethlehem is only a few kilometers to our northwest. Rachel, returning to and falling in love with her ancestral land, still remained an exile and an outsider. The titles she gave her three books are *Safiach*, meaning "the seeds that get left in the field after a harvest"; *MiNeged*, meaning "from the opposite side," and the posthumous *Nevo*, the mountain from which Moses was allowed to gaze at the Promised Land he was forbidden to enter, the mountain that towered so high in the minds of the settlers at Beit HaArava.

Rachel's spiritual descendants at Kibbutz Kinneret don't just work the fields anymore. They run a baptismal site.

You arrive there by taking a left off Route 90, just where the Jordan River sets out from the Sea of Galilee. A long lane, lined with willows and eucalyptus trees, follows the stream to a generous parking lot. It's nearly empty when I visit. October is off-season, and the current violent phase of the Israeli-Palestinian conflict keeps tourists away.

At the end of the parking area is a hewn, L-shaped limestone wall, inscribed with verses from the New Testament. At the bend in the L is a turnstile, manned by smiling, middle-aged, uniformed guards. As the information pamphlet notes, the visitors' center "has a wooden ceiling, huge pillars in the shape of a cross, a canal with water from the Jordan River and windows overlooking the river, designed with motifs taken from an ancient basilica." The snack bar "serves either the three-four buffet lunch or the traditional, well-known St. Peter's fish table lunch." Adjacent to the snack bar is a gift shop, where the good farmers of Kinneret sell Biblical Date Honey and Biblical Date Spread, as well as crosses and crucifixes for all confessions, vials of holy water, and stones from the Jordan. You can also rent a robe and a towel and proceed to the promenade above the river.

Steps lead down to the bank, where I spot a small group of high-spirited Americans wading in a kind of enclosure or pool set off from the main current by low metal railings and chains. A bearded man stands in the water and takes each of the white-clad devotees in turn. As I watch from above, he places a hand on a woman and lowers her into the green water, back first, until she is completely submerged. Her friends shout encouragement. As she surfaces, the rest, those waiting their turn and those who have emerged from the water, applaud.

I hesitate before descending the steps, fearing that I, a non-Christian, will ruin the spiritual moment. But they are friendly and welcoming.

"We're from California, members of a nondenominational Bible study group," a tall woman with sandy hair tells me as she dries herself off. The bearded man, she says, is their leader.

"What's the significance of what you are doing here?" I ask.

"The baptism? Well, this is where Jesus was baptized. It's very moving."

"But haven't you all been baptized before?"

"Of course. But it just feels right to do it here. It's so quiet and beautiful."

This shady, wooded stretch of the river, the only place where the water that flows within is really from Lake Kinneret, must certainly feel more spiritual to these Californians than the barren desert site down by the Dead Sea. "It looks just like what the place where Jesus was baptized *ought* to look like," another woman from the group declares. This is perhaps why they are willing to believe that Jesus was baptized here at Yardenit, even though the Gospels make it pretty clear that John the Baptist led Jesus into the Jordan's waters not far from Jerusalem. The Gospel according to John specifically says that that Jesus was baptized after he came from the Galilee to Judea.

But it's not my business to confront these nondenominational Protestants with inconsistencies. My religion has plenty of its own. People's mental pictures of biblical stories contain many details that are not in the text. What Jew can invoke his religion's founding father without seeing the boy Abraham shattering idols in his father's workshop? But that story is absent from Genesis. It appears only in later Rabbinic writings. Another example: Michael E. Stone, the Hebrew University's professor of Armenian studies, has written an entire book about the "cheirograph legend," a story that is very much part of the religious consciousness of many Eastern Christian communities. It tells how Satan deceived Adam and Eve at the time of their first sunset. The first man and first woman were in despair. "Now, when it became dark and evening arrived, likewise it seemed to them that darkness had come and there would not be light again . . . Therefore they wept and lamented till morning," relates one version of *The Cycle of Four Works*, an Armenian collection of apocryphal Adam and Eve stories.

"Now at daybreak Satan came in the form of an angel and he said to Adam, 'Why are you sad?'

"'But we do not know that we have done any evil such that God became angry at us and darkness seized us,'" Adam replies to Satan.

"Satan said, 'What will you give me if I give you the good news of light?' Adam said, 'If we could see the light again, we would become your servants, we and all our offspring.' Then Satan showed him the east and said, 'You will see the light there.'

"Adam and Eve looked to the east and saw that the sign of the sun's light had appeared. They rejoiced completely . . . Then Satan brought a stone and set it before Adam, and he said, 'Put your hand upon this stone and say thus, "Until the unbegotten is born and the immortal dies, all my offspring will be yours" . . . This was Adam's promissory note in Satan's hand. Now Satan brought the promissory note and buried it in the river Jordan. Again when the evening came and the sun set in the west, Adam knew that he had been deceived."

The promissory note is called the cheirograph. The story is uncanonical, and Stone says the texts that record it are never used in church. But no Armenian can recall the story of Adam and Eve without thinking of Satan's false bargain. Satan thought his conditions could never be fulfilled, but of course the unbegotten was born and the immortal did die. Armenian paintings of Jesus being baptized in the Jordan invariably depict him standing on and breaking a flat stone. It's the cheirograph, the promissory note, which Jesus voids. The story stands outside official theology, but it's more than just a story. It completely changes the significance of the exile from Eden.

Stone says: "According to the Bible story, Adam and Eve disobeyed God's command and were punished. But from within the inner circle of this story—and the Armenians lived within this inner circle—Adam and Eve were innocently deceived by Satan. That implies entirely different attitudes on sin and redemption."

When Yossi completes his own story, we all file slowly out of the cemetery. Like the boys and their teachers, I cleanse myself from the impurity of the dead by pouring water over my hands. I join them in the afternoon prayer service. The boys board their bus to continue their field trip, but two ask if they can get a ride with me. One has to catch a bus home for a doctor's appointment, and the other has a driving lesson in Tiberias. Their classmates head east, toward the Golan Heights, and we circle north, following the lake's shoreline. They listen politely when I respond to their query about the purpose of my trip.

"The rift valley is a natural object, created by physical forces. But when we look at it, we don't see just the physical object. We see stories and ideas and our own histories. People see the same landscape differently depending on who they are, when they live, what they've done, and what stories they heard when they were children. Any mere physical description of the rift will pale in comparison with how even the most literal-minded human being sees it," I say.

"For example?" says the one who has a doctor's appointment, whose name also happens to be Yossi, Joseph in English.

"Well, there are ancient Galilean legends that when the Children of Israel crossed into the Promised Land, after wandering forty years in the desert, they crossed here, over Lake Kinneret."

"That doesn't make sense," says the one who has a driving lesson. "The Torah says they crossed the Jordan at Gilgal. Near Jericho."

"And the legend says that when they crossed," I counter, "Miriam's well, the well that had given them water in the wilderness, remained in the lake, and that on a clear day you can see it, floating in the air."

"Miriam's well disappeared when Miriam died," Yossi objects. "That's why the Children of Israel started crying for water right afterward."

"The Torah doesn't say it disappeared."

"Well, the Rabbis do."

"The legend says that if you gaze on it, it cures you of your illnesses. Maybe you can save a trip to the doctor." I slow down. We've reached the city and its traffic. I let the boys off near the main intersection and continue north. It's afternoon, and the sun is descending behind the cliff face to my west. The water sparkles, but I see no well.

Route 90 turns into a snake north of the Kinneret, twisting and coiling up the steep ascent out of the graben. Nearly an hour after leaving the cemetery, I arrive in the town of Rosh Pina. When Rachel lived at Kinneret, this was a farming village, a small collection of stone houses perched halfway up a wadi on the side of Mount

Canaan. Now it's a collection of developments with down-to-earth houses that rent out rooms to travelers. I set about finding myself a place for the night.

Waking up the next morning on the slope of Mount Canaan is a familiar experience. For nine months in 1979 I woke up with the Hula Valley spread below my bedroom window and the sun rising above the Golan Heights beyond. I arrived there in January, after a three-month intensive Hebrew course in Kiryat Shmonah, half an hour's bus ride to the north. The fields in the valley and the scrub on the slopes were green then; red and yellow accents of poppies, anemones, and mustards appeared a month later in the wake of the rains. By June the sun had burned the land brown and flies droned dully in the heat.

On this October morning, twenty-five years later, there are clouds in the sky, but not enough for rain. A clutch of cranes flies overhead as I softly wheel my bicycle onto the road. I ride relentlessly uphill, until the road ends. From up here I can see Hatzor HaGelilit, where I lived for most of a year in 1979, spread out before me.

I'm on the other side of a divide. Hatzor, which abuts Rosh Pina to the north, began as a tent camp where the newborn Israeli state dumped some of the immigrants who flooded in during its first few years. The urgency of the moment and a painful lack of resources, combined with a large measure of condescension, led the government to build tent camps on available pieces of land. Fearing a crush in its central region and needing Jewish bodies to stake out a claim to more distant parts of the country, the planners saw the remoteness of the fields north of Rosh Pina—nearly a kilometer off Route 90—as an asset rather than a disadvantage. The new immigrants, the lion's share of them from Islamic lands, didn't see it that way. In fact, there was no economic or geographical justification for building a town here, except a vague hope that the established Jewish village of Rosh Pina, which dated back to Zionism's very earliest days, might somehow be a good influence.

By the time I took up residence in Hatzor, those original inhabitants who had pluck, business sense, or connections had long since fled to the big cities. They left behind several hundred families—most of them Moroccan, large, indigent, and unemployed. A small group of dedicated teachers and social service professionals from the center of the country did their best to combat delinquency and provide education. Hatzor had three major employers and a single tourist attraction. The employers were the municipality, the fruit canning factory on the edge of town, and the pipe factory in Kfar HaNasi, at the end of the road running east past the regional airport and Tuba.

The tourist attraction was a tomb on the mountainside above the town. According to local legend and an official sign erected by the Ministry of Religion, it was the final resting place of Honi HaMa'agel, a wonder-working rabbi of the end of the Second Temple period. The Talmud records two stories about him: how his obstinacy forced God to give rain in a time of drought, and how he slept for two generations and returned to find the land changed.

This time I'm staying on the other side of the divide, in Rosh Pina. My twilight search the previous night led me eventually to a room at Tamar and Arieh's. When Tamar sat me down at a metal-legged table in her narrow kitchen, I realized I was in the company of a type of Israeli whom one seldom meets anymore in the cities. We didn't talk politics, but these were old-time socialists. Their contentment with their small house and garden is one that derives, I think, not from their inability to afford or aspire to more but rather from a conviction that this is the proper way to live. To make extra money, they have erected two prefab cabins among the fruit trees in their backyard. A Russian family was staying in the larger one, and I received the smaller.

After my bike ride, I shower and walk down to the round gas station at the intersection. It used to stand on Route 90, alongside the local police station and a row of rather seedy stores and cheap restaurants. But more than a decade ago the main road was widened and rerouted to the east of the gas station, so that now drivers go by without seeing the old stores. A shopping center has gone up at the new intersection.

An Israel Antiquities Authority jeep pulls in, and after ascertaining that the driver is Yossi Stepansky, I climb into the passenger seat. He is Tali Gini's northern counterpart, a regional archaeologist in charge of protecting sites and artifacts. He also manages to do some original research and excavation.

Stepansky is round-faced and red-haired, and he wears a knitted kipah, as I do. He's got the air of an adolescent who's trying to restrain his enthusiasm in front of elders, which is odd because he knows a lot more than I do.

"So what do you want to see?" he asks me.

"I definitely want to see the Crusader fortress in Tiberias, and the excavations from the Talmudic period. Also, one of the early Islamic manor houses along the shore of the lake. And of course the Horns of Hittin, where Saladin defeated the Crusaders." I tick off these sites because I've been reading about them.

"But frankly," I add, "I know that the best things to see with an archaeologist are the things he's most excited about."

"Well, okay," Stepansky says, putting his hands on the steering wheel but not going anywhere. "Anything else?"

"A couple of weeks ago I read something intriguing," I say hesitantly. "It's an article by a historian named Elchanan Reiner, who goes to my synagogue. But I'm sure it's too speculative for archaeologists. You guys have your feet on the ground."

I pronounce that last sentence with a diminuendo, because when I mention Reiner, Stepansky's face lights up. He's already digging through the papers in his knapsack.

"You mean 'From Joshua to Jesus: The Transformation of a Biblical Story to a Local Myth'?" He pulls out a sheaf of photocopied pages. "It's a wonderful article. You said you used to live around here. Did anyone ever take you to Kherbet Shura?" And we're off. He swings the jeep out of the gas station to the Route 90 intersection. The light is green, and he goes straight through, onto the eastbound road to Tuba.

Halfway to the Bedouin village there's a factory, and we turn onto its access road. We overshoot the factory and head into the field behind it. We get out of the jeep and climb up a small mound. Large, dark hewn stones lay scattered among the thistles and weeds. In the Arava, structures are more evident because there is little vegetation to obscure them. Here you need to filter out the overgrowth before you see the facts on the ground.

Some of the rocks form a nearly square rectangle, clearly the bottom course of what was once a wall. On the inside margin of this rectangle there is another, somewhat lower rectangle. I know this pattern.

"A synagogue?" I guess. "Because of the benches."

"Definitely. It's oriented toward Jerusalem. But the Bedouin call it Kherbet Shura—the Ruin of the Council," says Stepansky. "When I first took up the post of regional archaeologist for the Antiquities Authority, my competence was called into question by a *mukhtar*— an elder—from Tuba. I came here to inspect these remains, which had never been documented before. The mukhtar knew from experience that places that catch the interest of archaeologists can be profitable to explore after the archaeologists are gone. So he came down from the village with a few other men to see what I was doing and offer assistance.

"'You of course know what this place is,' he said to me.

"'I know you call it Kherbet Shura. Why is that?' I asked him.

"'You mean you don't know what happened here, *ya* Yosef?'

"'I don't.'

"'With your name, you don't know? This is where Joseph's brothers held council to decide what to do with him after they threw him into the pit,' the mukhtar told me. 'The pit itself,' he said, 'is over there, at Jub Yusuf, a bit to the south and over the main road.'"

Now, anyone who knows his biblical geography knows that Joseph set out from Shechem, today's Nablus, to seek his brothers, who were grazing Jacob's flocks in the Dotan Valley, in western Samaria, part of the territory God granted to the tribe of Ephraim, Joseph's son. So how can it be that the mukhtar from Tuba knew for a fact that the events of the Joseph story all took place within a day's walk of his village?

The Bedouin of Tuba belong to the tribe of 'Arab al-Heib (the Bedouin word for Bedouin is "Arab"—non-nomadic Arabs are not, in the traditional Bedouin view, real Arabs). The Heib Bedouin are now just two generations past their nomadic origins and live in a few villages in the Galilee. Tuba is their eastern outpost. The tribe originated in the area around Baghdad. According to legend, its founder was a sheikh with seven sons who left Mesopotamia some three centuries ago and migrated westward in search of pasture for his flocks, just as Terach and Abraham did long before. My teenage acquaintance Naser and his friends had no interest at all in their tribal history and couldn't give me any more than these barest details. But like their ancestor, they looked west.

The common wisdom is that the local traditions that associate sites with biblical stories are products of the wishful thinking of Christian and Jewish pilgrims. On their visits to the Holy Land, they asked the locals where various events of the Bible occurred. The locals perceived the benefits that could result from having a holy site at or near their towns, so they obliged the foreign travelers.

Reiner argues, however, that many of these traditions predate the travelers. He sees evidence of an entire Galilean tradition that identified sites of biblical events in the north, even though the plain meaning of the biblical text places them in the center of the country.

"The Bible says explicitly where the story of Joseph and his brothers took place," Yossi says. "The brothers had taken Jacob's flocks to pasture in the Dotan Valley, in Samaria, northwest of Shechem. Jacob, whose tent is pitched outside of that city, sends Joseph to seek his brothers out. When he finds them in Dotan, they take him prisoner and throw him in a pit, and afterward sell him to a passing caravan on its way to Egypt. That makes sense because the Sea Road, one of the main routes to Egypt, runs along the coast not far from the Dotan Valley. So how did the Galileans transfer the story up to here?"

"It's not even on the road to Egypt," I note.

"Actually, it is. The medieval Damascus-Cairo road ran right by here." Stepansky takes a few steps to the north and spreads out his arms to show the road's direction. "That fits. But the Dotan Valley?"

"The landscape isn't right."

"But you have to see it through the Galileans' eyes. Because the hills and valleys here had different connotations for them." He looked west to Rosh Pina and above it to Safed. "Why is that mountain called Mount Canaan? They identified Safed with Bethel, the Israelite city in the land of the Tribe of Ephraim where Jeroboam, first king of the northern kingdom of the ten tribes, set up an altar and a calf to rival the Temple in Jerusalem. So the Dotan Valley could be here as well, alongside the Jordan River."

Reiner emphasizes that these Galilean traditions were not rivals of the biblical narrative. We have no evidence that the Jews of the Galilee and those who lived farther south rejected each other's traditions or possessed conflicting versions of the Bible. Rather, the Galileans read their sacred books in the context of the landscape around them, internalizing it as part of their day-to-day experience.

In his article, Reiner focuses not on the Joseph story but on Joshua. He quotes a German Jew, Petahiah of Regensburg, who visited the Holy Land in the latter part of the twelfth century, at about the time of Saladin's crushing defeat of the Crusader armies at the Horns of Hittin. "'And Mount Ga'ash is very high . . . And the mountain is fashioned with steps for ascending. Halfway up Joshua son of Nun is buried . . . Near them a spring of good water gushes from the mountain, and there are pleasant shrines built by the tombs . . . Near one shrine one can discern a single footprint, like that of a man walking in the snow, so it appears; it was here that the angel stood and the Land of Israel was agitated after the death of Joshua.'"

Mount Ga'ash, a set of twin volcanic peaks on the plateau of Mount Arbel, just above and west of Tiberias, is better known as the Horns of Hittin. The book of Joshua states clearly that Joshua was buried "at Timnath-serah in the hill country of Ephraim." The tribe of Ephraim's portion lay in Samaria, south of today's Palestinian city of Nablus. Nablus stands on the site of the biblical Shechem, the name Jews still use to refer to the city. The major tradition identifies Joshua's tomb in the village of Kafr Haris, south of Nablus.

The tomb that Petahiah refers to, halfway down the gorge below the horns, is now called the Tomb of Jethro and is holy to the Druze. According to Reiner, the tradition associating it with the Bible's Joshua begins to fade from the literary sources soon after Petahiah's visit. Other sources, however, identify the tomb with another Joshua, Joshua son of Perahiah, a sage of the later Second Temple period, at the beginning of the Common Era. This Joshua, or Jeshua, is a priest of Arbel, son of "the burned one." Reiner cites a Talmudic story about how Nebuchadnezzar, king of Babylonia, cast Joshua the High Priest into a fiery furnace along with two false prophets. The false prophets were burned, but Joshua emerged with only his clothes singed. In the story, this Joshua is a supernatural, messianic figure who can survive the fires of the Babylonian monarch who burned the Temple of Solomon.

Reiner practices the literary equivalent of what a geomorphologist does when he examines the rocks on a hilltop and determines that they were deposited in a riverbed. He adduces a rabbinic homily from the Jerusalem Talmud:

Rabbi Hiyya the Great and Rabbi Simeon son of Halafta walk together in the Plain of Arbel at daybreak. They see the morning star shining brightly. Says Rabbi Hiyya the Great to Rabbi Simeon son of Halafta: "Sir! So shall be Israel's redemption: first little by little; but as time passes, it shall gradually increase."

The Hebrew word here translated as "their redemption" is *ge'ulatan*. Reiner writes:

> The homily emerging from the dialogue of the two Galilean sages associates the future redemption of Israel with the Plain of Arbel. The association would at first sight seem arbitrary, but now that we have several highly charged references to that entire region, I venture to suggest—though still lacking documentary evidence—that the original version of this homily used the synonymous Hebrew word *yeshu'atan* rather than *ge'ulatan*, and that the latter replaced the former in the course of time, for reasons at which we can only guess. If so, it was not the mere rising of the morning star that inspired the homilist. Rather, it was the appearance of that heavenly body specifically in the Plain of Arbel, and the resulting association with the nouns Jeshua-Joshua-*yeshu'a*.

One of the Joshuas mentioned earlier, Joshua son of Perahiah, appears in the Babylonian Talmud as the teacher or companion of another, more famous Joshua—Yeshua, or Jesus, of Nazareth. Reiner also cites Jewish and Christian texts that associate Jesus with Tiberias and Arbel.

Mount Arbel looms before us as we descend south on the snake road to the Kinneret, its cliff face nearly vertical. It's not actually a mountain: it towers above the lake not because it is higher than the plain but because the land between it and the Golan Heights on the other side of the lake has collapsed into the rift valley. The lake, which lies before us like an impossibly smooth pasture, is wider in the north than it is down by the Kinneret cemetery. It looks like a teardrop. Yossi turns off the main road at a place called Karei Deshe, which means "meadow," and follows a dirt road up a mound.

"Most people think the lake is called Kinneret because it's shaped like a lyre, a *kinor*." We exit the jeep, and he helps me over a sagging line of barbed wire. "But the lake always gets its name from the area around it or the major city on its shores. Hence, the Sea of Galilee. Josephus, the Jewish-Roman historian, called it Lake Ginnosar, after the thriving area on its northwest shore at that time. Later, it was called Lake Tiberias. The Jewish sages of Tiberias were puzzled by the name, too. They debate it in the Talmud, tractate Megillah, page 6a, and come to the conclusion that the lake is named after the fruit of the kenar tree. According to one modern opinion, the kenar fruit is the artichoke. According to another, it is the jujube. But if you ask me, if it's called Kinneret, there must have been a town called Kinneret on its shores."

"At the south end. Kibbutz Kinneret," I recall.

"That's modern. The ancient Kinneret was here. That's why the Bible calls the lake Kinneret. Kinar was a Phoenician god, or perhaps a local one. That seems to be the source of the name." The Bible mentions it in a list of fortified cities in the territory of the tribe of Naphtali. Its name appears on a list, inscribed in the Temple of Karnak in Upper Egypt, as one of several important cities in the area conquered by Pharaoh Thutmose III in the fifteenth century B.C.E. That's the late Bronze Age, three centuries before Joshua.

We get back in the jeep and head south. When the road enters Tiberias, it turns into two congested lanes. We get stuck in creeping traffic, passing construction projects and restaurants. Yossi sees a pickup truck unloading laborers where a dirt path branches off the main road. "Excuse me, I hope you don't mind if I make a short stop here," he says, swerving onto the path. The construction site is adjacent to an ancient cemetery, and one of his jobs is to ensure that contractors don't destroy archaeological sites. "I was here the other day and gave the contractor very clear instructions about where they can work and where they can't."

He has a brief conversation with the Arab foreman, who assures him that they are carefully observing the boundaries the Antiquities Authority has established. The cemetery cannot be excavated because

Jewish religious law forbids graves to be disturbed. Ultra-Orthodox activists keep a careful watch on any site that they believe contains Jewish graves. Their criteria are not scientific, and many a rabbi has prevented the excavation of tombs that, the archaeological evidence makes clear, belong to other faiths, or to historical periods long preceding Judaism. Regarding the cemetery on Tiberias's north end, there is no doubt.

"This is the ancient Jewish cemetery of Tiberias," Yossi explains. "When we did salvage excavations here in 1996, we discovered a pair of mausolea from the Roman period. In one of them we found the inscription, in Greek: 'Joseph the son of Elazar, the son of Shila, man of Horsa.' Josephus mentions a Horsa 'on the edge of Edom,' where Herod met his brother Yosef on his flight from Judah to Masada in 40 B.C.E. So Joseph the son of Elazar, buried in Tiberias, is most likely from a family that came from the south. It's the one hard piece of archaeological evidence we have to prove that Jews from the south moved north after the Romans defeated them in the Bar Kochba Rebellion of 135 C.E."

Some scholars think that when Jews migrated from central Palestine to the Galilee after the rebellion, their stories and holy sites migrated with them. In suppressing the rebellion, the Romans largely wiped out the Jewish communities in the south, and the Galilee was left as the only part of the Holy Land where a significant number of Jews survived. The Sanhedrin, the religious court that was the Jewish population's highest institution, moved north to Sepphoris and then Tiberias.

The earliest documentary evidence of these traditions is found in lists of sacred sites compiled by local residents and travelers in the early medieval period. One of the first seems to date from the tenth century. But through analysis of allusions and stories in the Rabbinic literature, Reiner has come to the conclusion that at least some of these traditions were already current when the Jerusalem Talmud and the earliest Palestinian midrashim—Rabbinic homilies—were taking shape. That would have been as early as the third century C.E. However, he thinks, though he cannot prove, that they go back further than that and represent a tradition independent from and parallel to the one recorded in the Bible.

"Communities live their lives and see the world around them through the stories they tell. A story that took place far away, or in an inaccessible or unpleasant place, is not nearly so easy to make part of your life as one that takes place close at hand," I suggest to Yossi as we get back in the jeep. I tell him about my visit to the baptismal site at Yardenit.

"Let's not get caught up in speculation. We need to stick to the facts," he replies with a deliberate look on his face, like that of a boy who is determined not to stick his hand in the cookie jar despite his overwhelming desire to do so. A few months later, I saw him and Reiner in conversation at a conference in Jerusalem. They were looking at a photograph of a gathering of Safed Jews at Mount Meron before World War II, and Reiner was speaking authoritatively—I've seldom heard him speak in any other way—about the haircutting ceremony that, he was certain, was the occasion of the celebration. Reiner identified several figures and gave a short discourse on the history of the tradition of cutting a young boy's hair by the tomb in which, according to tradition, the Rabbinic mystic Rabbi Shimon bar Yochai is buried. Stepansky identified his grandfather. He listened politely but a bit impatiently, and after Reiner left, he said to me: "He gets too caught up in speculation. He needs to stick to the facts."

It's not a bad mantra for an aspiring scholar in this field, in which it's very easy to build entire theories of history around skimpy but suggestive physical evidence. Stepansky is enthralled by Reiner's work on Galilean traditions because it suggests that many Arab place names in the region are based on earlier Jewish traditions. I am also fascinated, because it means that those sites and names can connect us directly with the Rabbinic period, when the Jewish sages compiled the Talmud and other works of interpretation and legend that are now the foundation of Orthodox Judaism. The creed of today's Orthodox Jews assumes an unbroken line of tradition leading from the Pharisees of the Second Temple through the sages who lived and taught after its destruction and to the present day.

Modern archaeology and textual scholarship have cast doubt on the historical truth of Bible stories for decades now. The general public still seems to be shocked each time an article or popular book announces that there is no hard evidence to back up the stories of the Patriarchs, the Exodus, and Solomon's great kingdom. It's not news, however, to the scholarly community. A religious archaeologist such as Yossi Stepansky would be overjoyed if archaeological excavations corroborated these and other biblical stories, but with his training in scientific methodology, he has to admit the evidence is ambiguous, and he's learned how to live with that fact. He does so in part because in many ways modern Orthodox Judaism is not a religion of the Bible. It's much more a religion built by the sages upon the foundation of the Bible, after the destruction of the Temple and the rise of Christianity.

Many modern scholars now suggest that these sages were not a coherent group of scholars with a common worldview. Based on their analysis of Jewish and non-Jewish texts and on the archaeological record, they maintain that it took a few generations before rabbis, who initially operated largely independently of one another, coalesced into a school that set the religious agenda for the Jewish people as a whole. Reiner's work on Galilean traditions doesn't directly address this question; ironically, it offers a connection between place names we find today and the Jews of the Rabbinic period. But it also suggests that there were multiple narrative traditions at that time. Did the tradition that Yossi and I live by win out because it was the true one, or simply a better one, or was it just a matter of chance? If we were to turn the clock back and let it run again, might we find ourselves facing Tiberias when we pray?

"The Roman city was south of the present city, and only a small part of it has been excavated," Stepansky says as we continue into it. "The Ottoman city was here."

We turn south at the major intersection and confront the strip of dingy storefronts and hardscrabble falafel stands on the main tourist drag. In these tenuous times, there still aren't enough tourists to provide a livelihood for the locals. Foreign visits are up a bit now that terrorist attacks are down, and local tourism is also on the rise. Still, what we see looks bad.

"It's tough to do archaeology here." Yossi sighs. "Any equipment you leave out at night is gone by the morning."

"Here more than usual?"

"I'm convinced it's the climate. Here in the valley it's hot and the air is oppressive. So people who live in Tiberias are hot-tempered and have short fuses. In contrast, people who live in Safed, up in the mountains, are calm and easy to get along with." Yossi lives in Safed. "By the way, most of the Jews of Safed's old community lived in Tiberias first."

He pulls the jeep into the access road of the southernmost of the downtown beachside hotels. We get out and walk toward a cluster of old stone buildings. It's all that's left of the city's former Jewish quarter, or ghetto. This is where Yossi's grandmother grew up.

"She told me she saw Miriam's well. She knew how to swim. All the Jewish children in Tiberias knew how to swim, of course. My grandfather was from Safed. He didn't know how to swim. Her family came here in the migration of Eastern European Hassidic Jews at the end of the eighteenth century."

The Jewish quarter was cramped and, if we are to believe one visitor's report, squalid. James MacGregor, a Scottish adventurer, boated from the sources of the Jordan down to the Dead Sea in the 1870s, when the Jordan River was actually navigable. He made a stop in Tiberias. It wasn't devoid of culture; he was able to buy the *Times* of London. He also ran into some typical American tourists who were frantically trying to see everything worth seeing in the land of the Bible. "The great mistake they make is to go to many spots, and over many miles, rather than to see some places well," he wrote.

MacGregor liked Jews. In fact, he was something of a proto-Zionist. But he was clearly unnerved by what he saw in Tiberias's Jewish ghetto. And he apparently agreed with Stepansky's conjecture about climate and temperament: "The town of Tiberias is chiefly remarkable for the exceeding filthiness of most of its streets, and especially in the Jews' quarter. How any civilized European Jew can see his people degraded as they are in Tiberias, and then come back to his own gilded home in the west, and leave his brethren to wallow in such a mess beside that lovely lake, is beyond conception. Jews amongst us Gentiles in England have refinement, cleanliness, luxury, and elegance—why don't they send to the Rabbis of Galilee, at any rate, besoms and soap?"

Yossi doesn't want to talk about his family. He wants to talk about the excavation. He's so adamant about not mixing the two that I am hardly able to get anything out of him during our day together. Only later, when we meet in Jerusalem, would he agree to tell me about his own history.

"My grandfather was a Hassid of the Karlin court," he would tell me as we sat on a bench outside the conference room. "My grandmother's mother was a Heller—her great-uncle was the rabbi of Safed at the time of the great earthquake of 1837, when the city was leveled and thousands of Jews died. He helped save many of the wounded. They moved back to Tiberias, where my grandmother grew up. My grandfather left Safed after World War II. His aunt's husband ran off to the United States, leaving her with two children. She chased after him and took my grandfather and his older sister with her. The aunt ended up staying there. My grandfather came back to Tiberias, married my grandmother, and took her back to the U.S.—to Chicago.

"My parents came to Israel in 1950, as members of the religious Zionist youth movement Bnai Akiva. When I was four, we moved back to New York, and I grew up there. We frequently visited my grandfather and grandmother in Chicago." Yossi returned to Israel in 1969 to attend yeshiva and serve in the army; his parents and siblings followed a year later. He made his home in Safed in 1979, and he has lived there ever since.

We descend into a long, broad trench, a stone's throw from the Kinneret shore. Its walls are of layered rock, marl, and sand, and the floor is strewn with plastic soft drink bottles and other trash. We reach a spot where four courses of large building stones stand on either side of eight columns of paving stones. A white plastic chair stands forlornly to one side, just in front of a stone in the wall on the right side that is carved intricately with a border of undulating lines and balls, around a center section of leaves and a flower. On the left side there's a groove and a sort of foot projecting from the wall. Building debris and fraying, empty sacks lie around. Some construction workers are laboring on today's ground level, five feet above our heads.

"In 1948, after the War of Independence, the government razed the Jewish quarter of Tiberias to make room to build hotels," Yossi tells me, facing south. "It was a big mistake. They left only public buildings, like the synagogue of the Karlin Hassids that you see up there."

Now he turns to face north. "Two years, ago they wanted to develop this area, and we were called in to do a salvage excavation. We dug here for two seasons and made this discovery. We're standing in the gateway of the Crusader fortress of Tiberias."

Until recently, the scholarly consensus was that the Crusaders had built their castle away from the water, up on the slope. But in 1970 a historian named Zvi Razi took another look at the documentary and archaeological evidence and suggested that the fortress had been built on the shoreline. Soundings taken by an Israel Antiquities Authority archaeologist, Elliot Braun, also hinted that the fortress was in this area. The theory was disputed, though, until Stepansky uncovered the fortress's northern gate.

"The groove here is where the iron grille, the portcullis, was raised and lowered. The stone is typically Crusader—you see the smooth margins and the diagonal slashes. But they also took stones from what they found around them. This one with the wreath carved into it is just like ones we find in Roman and Byzantine synagogues. Maybe it was taken from the ruins of the synagogue that Jewish travelers from that period say they saw when they visited here, and which they called the 'synagogue that Joshua son of Nun built.' "

Tiberias, like most medieval Palestinian cities, was multicultural. The population bore the imprint of successive invasions, each bringing new customs. The Crusaders captured the city from the Muslims, who had in turn captured it from the Byzantines. It was founded in 19 C.E. by Herod Antipas, son of Herod the Great, who named it after his Roman imperial overlord. Local populations usually endured invasions and made do, so the city's inhabitants were members of all three great religions. They spent much time observing one another with incomprehension.

"The Franks are without any vestige of a sense of honor and jealousy," wrote Usama ibn Munqidh, "the gallant and cultured amir of Shaizar, whose life spanned almost the whole of the first century of the Crusades." I owe that description, and the translation, to Francesco Gabrieli, who has published a collection of Arab writings on the Crusaders and their times.

"Here is an example of this from my personal experience," Usama tells us.

While I was in Nablus I stayed with a man called Mu'izz, whose house served as an inn for Muslim travelers. Its windows overlooked the street. On the other side of the road lived a Frank who sold wine for the merchants; he would take a bottle of wine from one of them and publicize it, announcing that such-and-such a merchant had just opened a hogshead of it, and could be found at such-and-such a place by anyone wishing to buy some ". . . and I will give him the first right to the wine in this bottle."

Now this man returned home one day and found a man in bed with his wife. "What are you doing here with my wife?" he demanded. "I was tired," replied the man, "and so I came in to rest." "And how do you come to be in my bed?" "I found the bed made up, and lay down to sleep." "And this woman slept with you, I suppose?" "The bed," he replied, "is hers. How could I prevent her getting into her own bed?" "I swear if you do it again I shall take you to court!"—and this was his only reaction, the height of his outburst of jealousy!

Yossi points to a line of soot that runs along the western wall, a few inches above floor level. "There was a fire. But we don't know why or when."

This castle was the bait that Saladin used to catch the Crusaders and end nearly a century of Christian rule in Palestine. The Muslim general, who'd made himself first lord of Egypt and Syria, wanted to unite his territories and demonstrate that he could defeat the invaders from the West. He was also infuriated by Renaud of Châtillon's raids on Muslim pilgrims by the Red Sea.

Yossi explains: "The Crusaders and the Muslims played Ping-Pong for years, each invading the other's territory, moving their borders back and forth. That was until Renaud of Châtillon started harassing Muslim pilgrims on their way to Mecca. The final straw was when he set out from Aylah to raid the holy city. Saladin organized his forces and moved over the northern Jordan into Galilee."

The Crusaders' forces mustered at Sepphoris, seventeen miles away, just east of Galilean Bethlehem. Raymond III of Tripoli, ruler of the Galilee and lord of the fortress of Tiberias, reported to King Guy there. He left behind his wife, Princess Eschive, with a small garrison. Renaud of Châtillon had already arrived with his knights. The Christian force numbered about twelve hundred knights and

twenty thousand infantry, nearly every soldier the Crusader kingdom could call up.

Yossi leads me out of the trench and toward an old stone building right at the water's edge. "I want to show you something down
here. This is the old *mikveh*, the ritual bath of the Hassids." We walk
down a twisting staircase so narrow and steep that I have trouble
keeping my balance. There is very little light. We enter the cham-
ber where Yossi's great-grandmother probably immersed herself
monthly. Along the wall of the basin some plaster has been chipped
away, revealing a wall made of building stones with smooth margins and diagonal slash marks.

"I think it's a piece of the foundations of a tower from the castle.
No one noticed until now."

Yossi continues the story: "Saladin reaches Tiberias on July second. He breaks through the city wall, and the Crusaders take up positions in the fortress. That means there should be a wall around the
city, but we haven't found it yet. Lady Eschive manages to send a
messenger out to her husband at Sepphoris to inform him that she
is besieged. 'Come save us!' she pleads.

"Raymond tells the king: 'It's not to our advantage to go out to
battle now. They're setting us a trap. We can wait, we can negotiate.'
He sent a message back to his wife: 'If they take the castle, set out
on the lake with boats.' But Raymond wasn't on good terms with
the king."

The Knights Templar, who made up a large proportion of the
mounted men, convinced the king that it would be cowardly and
unchivalrous to leave a lady in distress. Guy ordered his troops to
march on the Muslims.

"The sources tell us that the Muslims entered the castle without
a fight and held the princess and her men hostage. But we have that
layer of soot," Yossi notes as we climb up the stairs into the sunlight.
"Some later sources say that they destroyed part of the castle after it
was already in their hands."

• • •

In his article, Elchanan Reiner refers to a Jewish biography of Jesus that was current during the period of Muslim and Crusader rule. A version of *The Chronicle of Jesus* has been pieced together by scholars from fragments found in the Cairo geniza—a synagogue storeroom for discarded documents—and elsewhere. It tells a very different story of the crucifixion than the Gospels do, and places it in Tiberias. According to this story, Jesus is arrested by the Roman emperor's men in the city by the lake and condemned to the cross. But Jesus uses sorcery to change himself into a bird. Joshua son of Perahiah, the Jewish sage, tells his student Judah the gardener to use God's secret name to turn himself into a bird and chase after Jesus, who has taken refuge in Elijah's cave on Mount Carmel. This is apparently not the Mount Carmel where the port city of Haifa now stands, for ancient sources refer to the hill above Tiberias as Mount Carmel and place Elijah's confrontation with the priests of Ba'al there. Judah captures Jesus, neutralizes his magic by wrapping him in a cloth, and hands him over to Joshua son of Perahiah in Tiberias. Joshua has his namesake crucified on a stalk of cabbage on Passover eve, and Judah buries him in an aqueduct in his garden. Seven days later, to counter rumors that Jesus has ascended to heaven, Judah exhumes his body and drags it through the streets of the city.

I have no way of knowing whether the person who wrote this, who was presumably a Jew, was trying to be funny or scurrilous. It's understandable that an early medieval Jew might have wanted to get back at the Christians through either parody or slander but not at all clear why he chose to set his story in Tiberias. Reiner doesn't have an explanation; he suggests merely that the stories of Jesus, Joshua, Jacob, and Joseph that we find set in and around Tiberias are all related to a native Galilean mythological tradition. But why is this city on a lake the site of Armageddons and redemptions?

Herod Antipas built the city to be his capital when he received the Galilee and a section of territory beyond the Jordan as his inheritance. The rabbis wouldn't live here at first, however, because the city was built over a cemetery. At a later date they purified the city and the Sanhedrin, the highest Jewish legal and self-governing institution, moved here from Sepphoris.

The center of the Roman-Byzantine city known to the rabbis lay to the south of the current city, and Yossi takes me to see the broad, rocky flats where it once stood, pressed between water and mountain. Yizhar Hirschfeld, the same archaeologist who argues that no Essenes lived at Qumran, is now excavating ancient Tiberias, though he won't resume work until November.

Yossi leads me over the barbed-wire fence that looks like no more than a symbolic boundary for the site. "The site was partially excavated in the 1950s, but the government decided in the 1990s to make its big investment at Beit She'an. The thought was that Tiberias had the beach and tourism, but Beit She'an had nothing else but history to draw visitors. Ancient Tiberias may be the largest and most interesting site in the country that hasn't yet been thoroughly excavated.

"There are a lot of puzzles here," he says, hands on his hips, his eyes squinting in the sunlight. "The written sources tell us the city was founded by Herod Antipas, who was a Roman vassal. But the excavators have found a good amount of pottery from the Hellenistic period, before the Romans came, which would indicate that there was already something here."

We look into a series of pits. "This was apparently the marketplace, near the typical roofed Roman *cardo*, with stores on either side. And the central bathhouse was here. But there's another opinion that it was a huge early Arab mosque. The locals of Tiberias call the cellar of the bathhouse 'the house of the midgets.' They've dismantled the mosaics from the bathhouse lobby."

There's another, larger pit, where I can make out a gateway and some pillars.

"This is the basilica. Yizhar claims that this was the seat of the Sanhedrin. It's similar to Roman period courthouses. These side rooms were where the tribunes sat."

"How can he prove that it's where the Sanhedrin convened?" I wonder. In my reading I've come across one of the puzzles of ancient Tiberias: whereas ancient historians, geographers, travelers, and rabbis tell us that the city was inhabited mostly by Jews, the artifacts archaeologists find indicate nothing about Jewish life. Without the writings, there would be little in the finds to suggest that the city was inhabited by Jews.

Yossi shrugs. "Good question. Lots of Roman courthouses have been excavated from this period, but no Sanhedrin seats. We don't know what one would look like and how, if at all, it would differ from a Roman structure. Yizhar's looking for something solid, like an inscription. One inscription can completely change the way you look at an ancient structure. Until then, he has to rely on his intuition and experience, and historical sources."

The object that excites Yossi the most is in the opposite direction, up against the slope of the mountain. He shows me several courses of stone blocks.

"When I first came here, there were only three stones sticking out of the debris, but I knew it was something serious. We didn't want the Tiberians to know about it for fear that they'd start taking it apart. I thought it might be a stadium, or perhaps the municipal building. I brought Yizhar to see it, and in the end they uncovered a monumental wall. Look how it curves. It's a theater, buried under a garbage dump. You wonder what shows played here. Maybe it was the first Jewish theater."

In one of the excavation pits alongside the wall, alternating gray and brown lines of sediment are off-kilter. The ones on the west side seem to have been pressed down a foot and a half by some violent force. The gray lines are full of pebbles and sand, the stuff that surging water brings in. This is a fault line, like the one I saw with Yoav Avni in the Negev. It is described in a geological report on the excavations by Shmuel Marco of Tel Aviv University and his colleagues: "Alluvial pebbles and fine-grained lake sediments, indicating a rise of water level, buried the Roman, Byzantine, and Umayyad buildings. The competent architects and builders of the Roman stadium were certainly aware of the seasonal fluctuations in the lake level. They built on the shore, confident that the structures were safe."

They conclude that the fault and the alluvium were the result of the great earthquake of January 18, 749. Contemporary writers tell us that the quake largely leveled Tiberias. They say that the city of Baalbek was completely swallowed up. A spring at Jericho reportedly moved six miles. A seiche hit the Dead Sea—its waters sloshed back and forth as if the entire salt lake had been sent sliding down the fault and then stopped short. A tsunami hit the Mediterranean coast, statues fell from their pedestals in Constantinople, and in distant Mesopotamia, a two-mile fissure opened in the ground.

Tiberias, like other cities along this crack in the earth, has suffered severe earthquake damage once every century or so since it was founded. There were major quakes in 30, 33, 115, 306, 363, 419, and 447. In 631 or 632, aftershocks continued for a month. In 1033 Tiberias was destroyed again in a series of earthquakes that lasted for forty days. Then came the quakes of 1182 and 1202. In 1546 a quake caused the Jordan to stop flowing for two days. A displacement of tectonic plates in 1759 killed between ten thousand and forty thousand people, many of them in a seiche on Lake Kinneret that battered down the walls of Tiberias. In 1837 the walls were destroyed again by a seiche; 28 percent of the population perished. That's when Yossi's great-grandfather helped dig people out of the rubble in Safed. In the twentieth century, 1927 and 1943 were the years of the great earthquakes. The twenty-first century has not had one yet, but it is coming. No wonder the people of Tiberias have short fuses. It's not the climate, it's the geology.

Yossi drives back into Tiberias, and we take the road that leads up from the graben, past sea level, to the plateau above. The city gives way to wheat fields. The remains of an old Roman road lie covered in the undergrowth alongside the asphalt. Yossi turns off the road and into the fields, following a dirt road, then pulls the jeep over to the side below a hill. We walk up.

It's not a hot day, but it's clear, and the sun floats naked in the sky. There are no trees, just grass and scrub. It's the last kind of place a soldier wants to be caught standing up in, and some old instincts are telling me to hit the ground.

There are two pinnacles at the top, set at diagonally opposed corners of a rounded square. Dark basalt boulders lie around the perimeter, and just inside it is a beaten loop path. Below us, to the west, is a broad field of wheat being tended by a farmer in a tractor. Beyond the field, on its north side, is the Arbel canyon. Trees are abundant on the slopes.

The two peaks are the horns, and the ancient town of Hittin lay below. The mountain is named after the village, as the lake is named after its city. The low circle between the peaks is a caldera, the mouth of a volcano that was spewing lava here on the edge of the rift not all that long ago in geological terms—beginning a million years ago and lasting until just a few tens of thousands of years ago, well into the period of human habitation. This is the Mount Ga'ash of which Reiner's medieval traveler spoke. The tomb of Joshua that the traveler saw is over on the slope of the canyon, by a spring. It's now the tomb of the prophet Shueib, holy to the Druze. They say that Shueib is Jethro, Moses's father-in-law, progenitor of the Druze. In Jewish tradition, Jethro is responsible for the first Jewish court system. Visiting the escaped slaves at Sinai, he found his son-in-law exhausted from judging each and every one of his people's legal cases, from lawsuit to spat. Jethro told him he must delegate authority by setting up lower courts.

"The Crusader force arrived from Sepphoris," Stepansky says, pointing down the highway that runs to our southwest. "The sun is setting, and Muslim sorties are beginning to engage them. The king makes his fatal mistake. He decides to make camp for the night in Maskaneh, over there, instead of by the spring."

It was a hot and humid night, but a day's hard ride north of the hottest spot in Asia. The next morning, Saladin's forces attacked.

"Saladin and the Muslims mounted their horses and advanced on the Franks. They too were mounted and the two armies came to blows," relates the Arab historian Ibn al-Athir, who had supported the other side in Saladin's wars to gain control of Syria. "The Franks were suffering badly from thirst and had lost confidence. The battle raged furiously, both sides putting up a tenacious resistance. The Muslim archers sent up clouds of arrows like thick swarms of locusts, killing many of the Frankish horses. The Franks, surrounding themselves with their infantry, tried to fight their way toward Tiberias in the hope of reaching water, but Saladin realized their objective and forestalled them, by planting his army in the way."

The Christian commanders sought to cut the Muslim lines. Raymond led his vanguard on a desperate charge, and they managed to break through.

"Raymond gets to the spring below, but he realizes that there's nothing he can do. His men and horses drink and begin to make their way northward to Tyre and Tripoli," Stepansky relates.

Ibn al-Athir quotes al-Malik al-Afdal, Saladin's son:

I was at my father Saladin's side during that battle, the first that I saw with my own eyes. The Frankish King had retreated to the hill with his band, and from there he led a furious charge against the Muslims facing him, forcing them back on my father. I saw that he was alarmed and distraught, and he tugged at his beard as he went forward crying: "Away with the Devil's Lie!" The Muslims turned to counter-attack and drove the Franks back up the hill. When I saw the Franks retreating before the Muslim onslaught I cried out for joy . . . This was how the tent fell: The Franks had been suffering terribly from thirst during that charge . . . but the way of escape was blocked. They dismounted and sat on the ground and the Muslims fell upon them, pulled down the King's tent and captured every one of them, including the King, his brother, and Prince Arnat [Renaud] of Karnak, Islam's most hated enemy.

Saladin was proud of his magnanimity. Ibn al-Athir tells us that, after the battle, the Muslim commander had his top-ranking prisoners brought to his tent.

"He had the King seated beside him and as he was half-dead with thirst gave him iced water to drink. The King drank and handed the rest to the Prince who also drank. Saladin said: 'This godless man did not have my permission to drink and will not save his life that way.'" Saladin meant that Muslim rules of hospitality required him to protect the life of any guest; by offering King Guy drink, Saladin had formally designated him as a guest—an honor he deliberately withheld from the man who had raided Muslim pilgrims in the south. "He turned on the Prince, casting his crimes in his teeth and enumerating his sins. Then he rose and with his own hand cut off the man's head."

A year later, Ibn al-Athir writes, he returned to the field of the dead volcano. "I crossed the battlefield and saw the land all covered with their bones, which could be seen even from a distance, lying in heaps or scattered around. These were what was left after all the rest had been carried away by storms or by the wild beasts of these hills and valleys."

"So the ground here must be full of artifacts," I say, gazing at a tractor driving back and forth in a field below us.

"Nothing," Yossi said. "I'd be happy to find even an arrowhead. But the land has been overgrown, worked and plowed for centuries. It's not like the Arava, where a traveler moves a stone from one place to another and that's where it stays."

. . .

The next morning, Thursday, I'm without a guide in unstable terrain. My plan is to visit a lapsed Muslim and a broken Christian stronghold, watch birds in a swamp, and end the day with devout Jewish friends. Before setting out on my trip, I reestablished contact with Naser, whom I saw last when he showed up for my wedding in Jerusalem in 1985. I had no address for him, and his full name was not of much use since nearly everyone in Tuba shares the same last name, Heib. Two decades ago phones were scarce up north, and many people, Naser's family included, didn't have one. The only phone number I can find is for the village school. When I give his name to the principal, I have to give identifying information as well, as there are other Naser Ahmad Kareem Heibs in Tuba. "He went off to study dentistry in Prague," I say. I expect to be told that he married a Czech and never returned. But the principal instructs his secretary to give me Naser's phone number. He's not only a dentist but also a doctor now, my old friend tells me happily when I call. He's got some business to take care of in the early part of the morning, so we agree that I'll come a little later, giving me some time to visit a couple of sites I never saw when I lived here.

When I leave Rosh Pina, it's pleasant and sunny, but in the few minutes it takes me to drive past the round gas station and across Route 90, the weather changes. A chilly breeze blows in from the northwest, and some gray clouds wander between me and the sun. Just beyond the Tuba turnoff, I take a left onto the packed-earth path that leads into the kibbutz's apple orchards. A cloudburst the previous night has dampened the earth, but it is still firm enough for me to drive on. After a couple of false turns I find the gravel road that snakes over the ridge and down the other side.

For a kilometer or so, the road runs just below and parallel to the ridge, behind the kibbutz. On my left, the rift side plunges toward the river, which flows through a canyon running more or less along the fault line. On the other side is the no less steep rise of the Golan Heights, a sparsely vegetated, dark gray mountainside. I encounter a white-bearded, retirement-age kibbutznik in shorts and earphones speed-walking along the road in the opposite direction.

"Is this the way to the Crusader castle?"

He looks vaguely down into the canyon and answers in the British accent of Kfar HaNasi's founding fathers. "People say it's down there somewhere. I've never been."

The road makes a hairpin turn and descends quickly into the gorge. Now I drive north, alongside the river. Even though the rainy season is just beginning, the Jordan is swift and lively, well-fed by underground springs that collect rain and winter snowfall from the surrounding highlands and mountains to the north. It is not comparable to most American and European rivers, but here, unlike farther south, it at least looks inviting. When I first met Naser, the riverbank was wild and not easily accessible. This is one reason why it was chosen as the location for a top-secret cemetery; Israel used to bury the bodies of terrorists and enemy combatants on the slope, not far from the fortress. Now there are picnic tables and rest spots among the willows, and hiking paths marked by the Society for the Protection of Nature.

I realize that more than an hour has passed, and I still haven't found the castle. I'm going to be late for my appointment with Naser, and my cell phone has no reception here. When I spot the broken keep, it catches me by surprise. It appears on a small hill between the road and the river, with a clearing below for visitors to park their cars. It has four names, depending on your mood. The Teutons called it Vadum Jacob, Jacob's Ford, since it stands guard over the ancient ford where caravans, armies, and nomads crossed the river on their journeys between Damascus and the sea. The French called it Chastellet, the little castle. The Arabs call it Beit al-Ahzan, the House of Grief. In modern Hebrew we call it Metzad Ateret, the castle of abundance.

Nineteenth-century travelers report a standing ruin, but the castle was later razed. Now, after excavation, its walls stand waist-high. Like all ruins, it's hard to fathom without a guide, but I can make out square corner towers that look much like those at the Roman fortresses at Yotvata. I wonder whether this demonstrates the persistence of good design or of a mythic connection the Crusader warriors may have felt with the Roman legionnaires of Diocletian's time. The Crusaders carved smooth margins around the edges of their building stones, just as Herod did.

This was the most convenient crossing place on the upper Jordan. Here the ancient Sea Road, connecting Egypt and the cities of the northern Levant, crossed the rift. It was therefore a strategic point on the frontier between the Crusader kingdom and the Muslim state, which was, at the time, consolidating under the leadership of Saladin. The ford was the military responsibility of the Knights Templar of Safed. In 1178 the Templars asked King Baldwin IV to allow them to build a stronghold to guard the ford, but the king refused. He had a previous agreement with his adversary, under which he undertook not to fortify the crossing. The Templars made their case again and again. Finally, in October, Baldwin gave his consent. In fact, he arrived at the site with his entire retinue and supervised the construction project himself. He persisted, even though Saladin offered him 100,000 gold dinars to desist.

The Muslims launched a series of attacks and a major offensive in August. With Saladin leading, they marched on the fortress. They dug a mine under the keep, filled it with wood, and set it afire. The Templar castellan cast himself into the fire. The Templars sued for peace, but Saladin refused. He killed eight hundred Crusaders and threw their bodies into a well, where they were discovered by the excavators. The Muslim forces destroyed the uncompleted castle. There were still Crusader tools scattered around the site, which have given archaeologists today a unique insight into the Franks' construction techniques. Three days later a plague broke out among the Muslims, and they evacuated.

They destroyed the fort, but the structure was doomed from the very start. The little hillock on which it was placed, so correct from a military point of view, sits directly on the central fault line. A series of tremors and major earthquakes has moved the walls and would have destroyed the outpost naturally. Its walls have been wrenched apart, and the section closest to the river has been pulled northward by a foot or so. Because the dates of construction and abandonment are known, geologists and archaeologists have been able to work together to map precisely how much the wall moved and when. The Crusaders unwittingly built a seismograph of stone that now serves as a prime piece of evidence for the Jordan rift valley being a strike-slip fault separating two tectonic plates.

Nearly a year later, I stand just below the castle, on the far bank of the river, with Naama Goren-Inbar of the Hebrew University's Institute of Archaeology. She kneels at the water's edge, the sinewy fingers of her right hand picking through the mud for bones and flint chips. Three bare-chested young men float lazily down the narrow river channel, laughing boisterously as they sprawl over a pair of overturned inflatable kayaks. Twin bailey bridges just to our north—the Daughters of Jacob Bridges—clatter like roller coasters each time a car passes overhead, so it's hard to hear my guide as she vents her frustration. "Wouldn't you know it. Now that we're here, I can't find a single bone."

On this small patch of riverbank, sheets of rock jut out of the ground and the river at odd angles. Wedged between the Jordan and an old Syrian minefield, it is no larger than a bedroom. Yet this spot, Goren-Inbar argues, can tell us whether our Stone Age ancestors from more than 780,000 years ago were able to cogitate, strategize, and perhaps even converse with one another.

In her archaeological publications, she refers to the site by its Hebrew name, Gesher Benot Ya'aqov (GBY for short). Seven excavation seasons and earlier surface surveys of the site have convinced Goren-Inbar that these ancient humans, of a now extinct species, were able to plan and carry out complex manufacturing processes, hunt and dismember large game, and control fire.

Unfortunately, she has found no remains of the people themselves. But there are abundant bones of animals who shared this lush Eden with them, including the skull of an elephant that they hunted. The tools they made—which archaeologists classify as hand axes, cleavers, and pitted stones that may be kinds of nutcrackers— are finely crafted, despite their rough appearance.

A stone-working hominin species was the first to leave Africa and spread into Asia and Europe. The rift valley, lower than the plateaus on either side and lined with springs, streams, and lakes, provided a lush and warm corridor northward, where grasses and game animals abounded. During the 2 million years before humans arrived, the opening of the rift was associated with spurts of volcanism. The basalt rocks of the Golan Heights are hardened lava from those volcanoes. Lava flows created basalt dikes in the valley, which worked like an underground dam. They prevented the flow of water from the basin to the north of us, the Hula Valley, to the much deeper basin of Lake Kinneret to our south. The backed-up water formed a lake.

"That's why it's so flat," she says, pointing at the checkered farm fields of the Hula Valley. "It's a lake bottom."

The lake was fed by melting mountain snows and springs, so it was always fresh. Here, where the wide graben turns into the narrow fissure of the fault line between the great plates, the lake extended a narrow finger. Humans lived along its shore.

After World War I, when the British took control of Palestine, the river became a border between the British Mandate in Palestine and French Syria. The British built a customs house and stationed a cavalry unit nearby to protect the crossing.

Goren-Inbar waves her hand in the direction of the bridges. "In 1934 or 1935, the British decided to enlarge the bridge—an old one, not the ones you see now. They dug down and found hand axes." British paleontologists had already unearthed remains of prehistoric humans in Palestine. Dorothy Garrod, a pioneering archaeologist who would later be appointed Cambridge University's first female professor, was excavating early human remains in the caves of Mount Carmel at the time. The cavalry officer had the finds sent to her, and she confirmed their importance.

When, in 1948, the river became a hostile border between Israel and Syria, further work became impossible. Interest in the site resumed after Israel's conquest of the Golan Heights. Goren-Inbar came here almost by chance, as the result of a request to help a colleague who had noticed that the castle's glacis—its outer embankment—was made of lake sediments. He thought Goren-Inbar, who had dug early human artifacts in Africa and at Ubeidiya, might be able to help date the sediments.

"I said absolutely not," she recalls. She believed that the lake bed stratigraphy offered no firm clues. "But I took a day off and walked below the castle with a geological hammer to have a look. I found a hand ax and several cleavers. It showed where the lakeshore had been. That was the first indication that the site was much longer, and it gave me a feeling that south of the bridge would be a better place to dig."

The stone tools Goren-Inbar and her predecessors found at the site are bifacial—they have been banged and chipped on two sides to create sharp edges. Most of those in the more ancient strata were made not of flint, the stone favored in later periods for stone tools, but of basalt. Bifacial, basalt stone tools are also prominent at sites in East Africa. The culture that produced these tools is called Acheulean, the culture that the first human emigrants from Africa took with them into the Levant. The earliest Acheulean finds in Africa date to about 1.6 million years ago. The earliest ones in the Jordan Valley—those at Ubeidiya—are dated to 1.4 million years ago. In geological terms, that is an instant. It means that the Acheuleans radiated out of Africa almost as soon as they appeared.

"These early hunter-gatherers had big brains and mental abilities that seem to have been much higher than those of their predecessors," Goren-Inbar had explained to me during an earlier visit to her office. Short, dark, and intense, she jumped from desk to whiteboard to bookshelf as she talked. "This is the hominin we find in Java. We don't know the details of his voyages because we're lacking evidence. Many parts of Eurasia don't have an abundance of the volcanic material that helps us date artifacts. Hominins were tied to freshwater sources because they still didn't have ceramics to store and transport water in." But the material they needed for their tools was not available nearby. True, the people of Gesher Benot Ya'aqov could reach out and touch the basalt boulders of the Golan Heights, but the nearby rocks are rough-grained and porous, inappropriate for tools. They had to go far afield to find the high-quality, dense basalt they needed.

"That means they had to map the formations, schlep tools, search for raw materials. They had to know where to go, how to cut it out of the large block of stone, and where to bring it for further processing," she explained in her office. And on the site: "These people were very sophisticated. They could identify what results different technologies would give them when applied to different stones. Everyone thinks Stone Age humans were like this"—here she raises her arms like claws, hunches her back, furls her brow, and thrusts out her lower jaw, drawing glances from more rafters—"but they had big brains."

The most dramatic evidence that these lakeside people were capable of planning and strategy are the remains of what Goren-Inbar interprets as an elephant hunt. In 1989 she and her fellow excavators unearthed the overturned skull of an elephant. Under the skull was one end of a long, thick oak log. Close by was a basalt core, a large rock cut to form a sharp working edge. Based on the placement of the artifacts, Goren-Inbar and her colleagues believe that the shore dwellers killed the elephant, then used the log to overturn its skull and the basalt core to crack it open. Brain was a prized cut of meat.

The question of whether early humans actively hunted large game or simply scavenged beasts who had died of other causes has been a lively subject of debate among scholars. The Gesher Benot Ya'aqov elephant provides strong evidence for hunting. Although many scholars think language emerged much later, Goren-Inbar finds it hard to comprehend how such a complex operation could be planned and carried out if the hunters in the band could not communicate with one another. Unfortunately, language does not fossilize, so both sides of the debate can cite only indirect evidence.

The elephant hunt makes for a good story, but how can we know it happened? An inch of sediment represents thousands of years; we can't be absolutely sure that the log and the core and the skull were all left there on the same day. We can't even be certain why the people at the lake would have wanted to hunt an elephant. Studies have pointed out that large game hunting seems to make little sense in terms of diet. An elephant contains much more protein than a small hunter-gatherer band can safely consume in the time before it spoils; in terms of effort expended versus nourishment attained, it would have made a lot more sense for the Acheuleans to have hunted rodents or scrounged for grubs. Some archaeologists have recently proposed that early men didn't hunt large game for food at all. They did it, these archaeologists suggest, to gain status and impress early women—a theory that would please Amotz Zahavi.

When Goren-Inbar argues that these humans could not have planned and carried out the hunt without language, she knows she is going out on a limb. The problem is that no one really knows how cognitive ability links up with complex actions such as hunting. Lions and hyenas hunt in packs without language. Ants cooperate to build complex nests with only chemical cues to guide them. Burials and cave art provide evidence of the human use of abstract symbols, but these appear in the archaeological record hundreds of thousands of years after the putative elephant hunt at Gesher Benot Ya'aqov. Our own language has become so inseparable from our thought processes and worldview that we can't conceive of people interacting in its absence. But that doesn't mean that here, nearly 800,000 years ago, people did not live without syntax and grammar.

Goren-Inbar insists on climbing down on a slippery rock ledge that juts out at the water's edge. The moving tectonic plates have crushed and bent the strata here so that they approach the vertical, but in different directions. We and the Acheuleans literally live on different planes. The result is that, when you dig down, you don't necessarily go from more recent to less recent finds. On the patch where we stand, the layers go from north (younger) to south (older) strata.

"You see those ripples in the river there? That's caused by rocks sticking up from the riverbed. It shows where the next stratum begins," Goren-Inbar tells me.

All the rich finds come from an amazingly small patch of riverbank, no larger than a modest suburban backyard. Despite its hugely important scientific value, the site has suffered man-made destruction time and time again. The strata Goren-Inbar excavated represent a window of about half a million years of human habitation. She believes that the site was inhabited for twice that time. Yet archaeologists may never be able to study remains from periods earlier or later than what she has found. Turkish and British bridge-building and flood-control efforts included digging and earthmoving around the channel. Israel's drainage of the Hula marshes in the 1950s caused even further damage.

"When they drained the swamp, the sediment, which had been waterlogged, started compacting. As a result, the floor of the valley subsided," she explains. "Soon the river channel, which sat on basalt, was higher than the valley and started flowing back into the Hula. So they sent in equipment to deepen the channel."

The bulldozers shaved some six meters of soil off the riverbank and dumped it farther up the sides. Those six meters represented hundreds of thousands of years of human lives. The refuse piled up on the hill is rich in hand axes and other artifacts, but they cannot be dated, and their original context cannot be documented.

The latest wound to the site came in 1999, when the government body in charge of water management in the Hula Valley sent in heavy earthmoving equipment under cover of night—violating Israel's Antiquities Law and previous agreements with the Israel Antiquities Authority. The action caused huge and irrevocable damage. Among other things, the dredging of the riverbed has made it impossible to excavate the site's oldest strata, those predating the earliest ones Goren-Inbar's team has excavated. Most of these strata lay under the river channel, until it was plowed up and dumped along the riverside.

Bent over at the waist like a rice farmer, Goren-Inbar searches through the mud. "Here's a bone fragment, and here's a flint," she says to me from behind her legs. She clearly wants to find something more impressive for me. A truck thunders over the bridge, heading for Route 90.

When I arrive in Tuba, I find it transformed. The community now extends to the road itself. In the place of the old corrugated aluminum shacks, there are large houses, a boutique, and a travel agency. The village's new prosperity may be a product of the strong economy of the 1990s. It also owes something to Jewish generosity: the Jewish Community Federation of San Francisco has adopted the village and sponsors a number of philanthropic projects there.

Naser greets me outside his clinic, which is on the ground floor of a two-story house, just past the second of two new traffic circles along the central road. He awaits me with a broad smile, next to the sign that declares "Heib Naser Karim, MD, DDS." By my calculation he must have hit forty recently; his wiry body has filled out, and his unruly hair is thinner. He's also got a softer, mellower expression than that of the discontented adolescent I knew a quarter century ago.

Proudly, he shows me the clinic: a waiting room, an examination room, a consulting room, and a dentist's room. He works here single-handedly, without a secretary, nurse, or hygienist. His dentistry is all private, but the medical care is mixed—most of it comes through the health plan that the majority of the village's residents belong to. He's got competition from another village doctor.

His house lies a short drive away, behind the one belonging to his parents where he hosted me on my first visit. It's low-lying, impeccably clean, and there are no mattresses in the living room, only the standard couch and armchairs. His two young daughters come running to meet him, and his wife comes to say hello, then brings coffee and cookies. It's Ramadan, the Muslim holy month in which believers fast during the daylight hours, but Naser remains unobservant. We update each other on our families and our careers.

"When you went to Prague, you said you'd never come back," I remind him.

He shrugs. "I could have stayed. I had a girlfriend there. But after a while I decided it wasn't fair to my parents. They paid for my education. I needed to consider their feelings. So I came back."

I grin. "And you said you'd never get married and have children."

"I really thought that. For a long time I didn't. But then something changed. She's from Haifa. I met her at a wedding. We got married five years ago. We go every Wednesday to visit her parents there. I would have invited you to the wedding, but I didn't know how to get in touch."

We talk about the week's events. He favors disengagement from Gaza and thinks that the Israeli government needs to move decisively toward reaching an agreement with the Palestinians.

Like his forebears, however, Naser looks to the west, to Haifa, and to the surrounding Jewish settlements where he has patients. He's puzzled when I tell him about my visit to the Crusader castle.

"I've still never been," he says.

As I head west toward Route 90, I drive into the wind. Beyond the highlands before me is an unseen sea. Streams of air sweeping down from northern Europe skim its surface and pick up moisture, which they release as they run into the stony barriers pushed up by the rift. Some fall before the rift, some within it, as the airborne sea crashes into the rift's eastern wall. In the wilderness, God promised the Children of Israel that in the land He had chosen for them they would not have to labor to bring water to their fields, as the farmers along the Nile had to do. In their new home, He told them, water falls from the heavens. He accomplished this miracle by exhausting the sea winds. After their collision with the mountains of Bashan, Gilead, Ammon, Moab, and Edom, the winds are battered and are forced to drop their cargo. Lighter, they rise and sweep out over the Transjordanian plateau, but they have little to give to the desert beyond.

I turn north. Clouds are coming in, low-lying and gray. These are not the ominous black thunderclouds I knew in my American childhood, which looked like the exhaust of diabolical furnaces. We seldom get such invaders over here. Levantine clouds are more airy, less electric.

My turnoff appears a few miles to the north. The asphalt lane is lined with eucalyptus trees; the road is littered with sheets of their bark. There's a visitors' center where you can rent carts and bicycles for a ride around the lake. Ronny is waiting for me there.

He's dressed in his usual sandals, blue work shorts, and faded T-shirt, and he's got a camera hanging from his neck.

"I didn't catch before what your connection to this place is," I say to him as he leads me toward a jeep that's waiting a few yards down a broad, white gravel road.

"I do some work here, photography and stuff." He peers into the jeep. "Meet Nadav."

When the early Zionist pioneers arrived here at the end of the nineteenth century to farm the fields of Yesod HaMa'alah, the land was ominous and dangerous. Slimy things slithered through the mud, and the air was infested with millions of insects, among them the dreaded anopheles mosquito, which carries deadly malarial plasmodia. Marsh Arabs were living specimens of what the swamp could do to the human body—they were stunted, muddy, ever-hungry, and all but the most resilient of their children died very young.

Rachel the Poetess loved the Kinneret's south shore; she sang of the trees and flowers and waves on the water. But the socialists who founded the collective farms had absorbed from Marxism a conviction that they could remake and control nature as well as human beings. Having achieved their Jewish state, they had a hungry and homeless population of refugees to feed and clothe. The Hula Valley was useless to men so long as the dike of solidified lava at its southern end kept the Jordan's waters from reaching the lake below. So channels were dug and the marshland was drained. The former islets and coves became fields of wheat, cotton, and vegetables.

But nature acted the reactionary. The collective farmers found that the reclaimed land was not all that fertile, and they had to leave much of it fallow. With its water siphoned off, the peat soil began to dry out. Sometimes the ghost of the old swamp would appear in the form of spontaneous underground conflagrations. In the meantime, green won out over red. Socialism waned, individualism triumphed, and Israelis gained a new appreciation for wetland ecologies. The government decided to reclaim a part of the fields for a reconstituted marsh. The snakes and centipedes were invited back, as were the mosquitoes, anopheles included.

"I'm really sorry, but I've got to get going. I've got a meeting down south," Ronny says. "But Nadav will show you around."

Nadav is a lithe man of thirty. He lives at Ayelet HaShachar, the kibbutz across the road from the ruins of Hatzor—not the modern town but the tell of the powerful ancient city after which it is named.

"I used to work in the fishponds," he explains. "But I got more interested in the birds that ate the fish. So when they opened the swamp to the public, I applied for a job here."

He eases down the hand brake, and the jeep begins a clockwise circuit of the long, low-lying lake, whose undulating shore forms a series of bays and inlets. Here and there islands interrupt the barely stirring surface of the water, which is thickly thatched with yellow grasses. Reeds grow profusely along the water's edge and in patches farther in. Under the gray sky, the Naftali highlands, the mountains bordering the west side of this part of the rift, look like a distant projection of the lake onto a cathode-ray screen. Small, squawking congregations of white cranes dot the water. Occasionally, one bird stretches out its black-edged wings, as if it has decided to seek more piquant gossip elsewhere, but then it adjusts its black legs, folds up its wings, and remains. Some water buffalo wade into an inlet as we pass by. A cat-size rodent lumbers across the road, glares at us briefly, and dives smoothly into the water.

Nadav stops so that I can take a look at the apathetic ruminants. "We brought a tribe of water buffalo from Sudan to reestablish the native population. The nutrias are another story. They're South American, not part of the original fauna at all. Some kibbutzim imported them years ago and tried to raise them for their fur. It didn't work, and some escaped into the wild and established themselves in the swamp. They're doing very well. But the thing that most interests me is the birds."

The cranes are most visible, but a second look reveals northern shovelers and mallards paddling in the lake. A white-breasted, red-beaked kingfisher shoots off a perch on a wooden observation post, snatches something out of the water, and flies off to the other side.

"Israel has the highest density of bird species in the world," Nadav says. "There are 530 species that live in or migrate through the country on an annual basis. The rift valley is the major migration route for birds that summer in Europe and winter in Africa and southern Asia." The valley produces thermal convection currents as winds come in from the sea and literally fall into the rift. Birds minimize their energy expense by riding these currents.

"I've got to go back and guide some visitors. But you should stick around. The cranes are feeding now in the nature reserve just south of here, but now's the time when they come back to the lake for the night. It's a sight you shouldn't miss."

I hitch a ride back to the entrance with him, unload my bike from my car, and take it for a swing around the lake. At the spot where I watched with Nadav, I stop and sit at a picnic table near the water's edge. Where the sun descends over the Naftali heights to the west, the clouds glow weakly, as if they cannot withstand the insistent wind that blows in from the plateau and swoops down over the valley. It is the roosting hour. White pelicans have already taken up positions in the shallows. Mallards and northern shovelers continue to feed placidly, but the cranes bleat their evening song, an alto trochee, high then low.

Bands of snake-necked birds begin to fly in from the nature preserve to the south, seven or eight at a time, gliding over the manmade marshland, looking for company. Most head to the north end of the lake, where many others of their kind already stand up to their knees in water. A few, however, descend in an inlet close by me, a little beyond where the ducks swim. I feel a drop; there's a mist to the north that might be rainfall coming in. A nutria swims by in the inlet to my left, climbs out, and skulks off. Another, much larger crowd of cranes circles in from the west; a few independent-minded birds break off and head my way. More raindrops, as the wind shifts to the northwest; the reeds begin to bow in my direction.

Another flock comes in, two hundred at least. They fly straight into the wind, so when a gust shoots down from the north, they float nearly motionless, an illusion of imperviousness to the forces buffeting and pulling at them from above and below and all sides. They seem so detached from the physical world that, for the first time, I really understand why the Holy Spirit is represented as a bird.

A thousand cranes now circle in from the west. Another mass rises up from the lake and circles behind them, forming two quadrillelike ranks moving in opposite directions. Off to the side, two social outcasts flee desperately from three adolescent toughs. If I'm anthropomorphizing the birds, I'm in good company. After all, Naama Goren-Inbar anthropomorphizes ancestral humans when she argues that they could speak. And the coldest of scientific minds sometimes ascribe desires, wills, and purposes to physical objects, whether animate or inanimate. And when we speak of other human beings, we inevitably color them with our own fears and aspirations.

The birds in the huge assemblage emit soft sounds. They have a hard time against the wind. One has a leg sticking out; perhaps he's wounded. The flock circles and lands, wave after wave, in the water for the night. They are indifferent to the pounding rain. For them, it has no meaning. I, in contrast, must find a dry and warm place to sleep.

Dark clouds hang low over Mount Hermon when I mount my bike in Kiryat Shmonah early Friday morning. I can get a ride in before I go because the storm has paused for a breath. But rainwater runs along the sides of the patched asphalt street that zigzags down from the Menara cliff, where the ruins of a Crusader castle lie. Time and fear of lightning keep my ride short. I reach a spot called the Geological Garden, a bare cliff face that reveals the strata of earlier eons. It's a favorite rappelling and climbing spot. Earthquakes have split rocks and sent them tumbling below, where they've been covered and ground into silt and soil by earlier incarnations of today's rivulets. Kiryat Shmonah, Israel's northernmost city, is built on top of the rockfall, on top of the ruins of the Arab village that preceded it, and on top of even older villages, pastures, cairns, and roads.

Kiryat Shmonah was my first home in Israel. I arrived by night, so my initial view of the country that would become my home was of the basin of the rift valley, laid out in newly planted squares of grain. The thick, gray limestone slabs of Mount Hermon, formed at the bottom of an ancient sea, towered to the east over the dark basalt Golan plateau. Later, as a soldier, I manned outposts on the peaks of that mountain and also at the edge of Lebanon's Bekaa Valley, another graben in the progress of the rift before it reaches its northern limit in the mountains of Anatolia.

Even though a quarter century has gone by since that earlier October, visiting Kiryat Shmonah is like coming home. I spent the night in the bomb shelter qua ground-floor vacation apartment that my old friends Yossi and Havva rent out to vacationers. Just two years older than I, but already parents of two children when they unofficially adopted me back then, they are now grandparents several times over, while their youngest children are still in elementary school. From them I learned the importance and pleasure of careful religious observance within a community, although I rejected their politics. It's a delicate time between right and left in Israel these days, and we have different interpretations. Yossi and I skirt the subject of the Gaza disengagement plan as we update each other, over some late-night sandwiches, on the progress of our children. Havva arrives after I go to bed—she has spent the day at a marathon meeting in Jerusalem about how to transfer the Jewish children of the Gaza Strip to other schools when the country dismantles the settlements they live in.

I leave without saying goodbye. Havva is sleeping in, and Yossi has to get the kids up and out to school alone. As I head down to Route 90, the storm lets out its breath and a drizzle begins. Half an hour later, as I pass Rosh Pina, the drops are as big as pearls, and a half hour after that, as I drive through Tiberias, the drops have merged into sheets of water. The narrow section of Route 90 that runs along the lake's west shore is clogged with traffic. Rachel the Poetess's grave stands lonely on its windswept tell. I pick up speed after the road turns south, past Sha'ar HaGolan, Beit She'an, and Tirat Tzvi. I slow down for inspection by the poncho-clad soldiers at the crossing point that marks the boundary between Israel proper and the West Bank. Two young yeshiva boys in T-shirts huddle under the checkpoint's corrugated metal awning, and I offer them a ride.

They want to listen to the radio. The always inferior reception of the Jordan rift valley is even worse because of the storm. We catch snatches of a call-in program: angry listeners call to vilify Prime

Minister Sharon for his plan to withdraw from Gaza, or for not withdrawing from all the occupied territories at the same time. Between bursts of static, I learn that the two boys have just enrolled at a new yeshiva in the north, one for military-age kids. But because it is designated a "higher yeshiva," the boys are in fact exempt from the army, although according to the yeshiva's program they are to do minimal service of six months. They say they are not sure they will serve even that, given that the government has abandoned its commitment to holding and settling all of the biblical Land of Israel. One of them lives in a religious neighborhood in Jerusalem, the other in a West Bank settlement not far from the capital. They represent a worrying trend of separatism in the modern religious Zionist community. Families and communities that once viewed service to the State of Israel, especially military service, as an integral part of their religious philosophy, now seek to isolate themselves in small communities of true believers. Having ceased to further their messianic theology, the state, they believe, no longer has a claim on their loyalty.

I take the twists and turns of the northern Jordan rift slowly. Every so often the headlights of another vehicle appear, very close, coming from the other direction. Then the road straightens, and we reach the junction where the Tirza River meets the Jordan Valley. The river, usually dry or a trickle, is a torrent. It has broken free from its narrow channel, and fifty meters of road are covered with water. It's impossible to tell how deep the water is or how strong the current, but I see other cars inching their way across. The radio signal finds us long enough for us to hear a report about the rescue of passengers from a tourist bus caught in a flash flood near the Dead Sea hotels. I look at my companions, and they shrug. The river doesn't seem to be sweeping any of the other cars away, so I join the convoy and ford the stream in first gear.

We brave two other, smaller floods before reaching the next big canyon, the Petza'el riverbed, the junction with the road that ascends to Gitit and the other Alon road settlements. The Alon road, going north, links up to the settlement where one of the boys lives and thence to Jerusalem's northern neighborhoods, so they ask to be let off.

At the junction, where I expect a river flowing to the Jordan, Route 90 is more or less passable. The road is not underwater, even if the atmosphere is. I find the river two minutes later. It has diverted itself into the fields of the adjacent Jewish settlement, which now look like rice paddies under a fierce monsoon. A huge section of road is submerged, and some cars are turning back. Emboldened by the three other floods I have forded, I press on. At the deepest point, the water reaches halfway up the wheels, but a look in the rearview mirror convinces me that the far shore is now closer than my embarkation point.

As I pass Naama, the rain seems to let up a little, but that's only a feint. When I reach Uja, the storm resumes in full force. The village boys and men are out in the rain, milling about or heading south, and toward the southern end of the village I find myself backed up in traffic. Cars ahead of me make U-turns, and I move forward in line until I reach the lip of the Uja river channel.

Villagers stand on the muddy knolls that flank the road; truck drivers and travelers have pulled their vehicles over to the shoulders and climbed out to take in the view. The Uja, usually a dry depression, has become a primal torrent, toppling boulders and carrying stones leagues from their points of origin, like the ancient Paran. The river has torn away the guardrails, and rocks carried down from the hills to the west have landed on the asphalt, producing a line of sinister rapids. To the east, where the channel narrows as it approaches the Jordan, the river flows fiercely but within confines; to the west, however, the entire landscape, as far as the eye can see in the storm, is underwater. The deluge plummets down from the mountains and wells up from the depths of the earth. The mosaic of the landscape has been flushed. There are no longer any facts on the ground.

An oil tanker tries to make the crossing, but the driver halts just a minute after making contact with the water; village men and boys frantically wave their arms to convince him to stop. A gaunt, elderly man of about sixty shakes his head and shrugs his shoulders when I meet his eyes.

"Does this happen every time it rains?" I ask. In my quarter century of traveling the rift valley, I don't remember ever having encountered a flood here.

"I've never seen anything like it," he says.

I cannot go on. I turn back through Uja and wade again through the Petza'el ford. At the junction where I let off the two yeshiva boys, I turn west. As I ascend the wall of the rift, I leave the heaviest rain behind me. When I turn south on the Alon road, the precipitation is of unmythical proportions. I notice that my gas is very low, but I figure I can make it to Jerusalem. A half hour later I realize I must have miscalculated, but this stretch of the higher road has no sign of human habitation, no gas stations, and virtually no other travelers. The road seems like a perfect spot for terrorists from the villages invisible behind the ridge to my west to ambush a driver who has been so idiotic as to get here with insufficient fuel. I drive for the next half hour on the verge of panic, but then a gas station sign comes into view, towering over an Israeli settlement of red-shingled houses.

It turns out to be a simple self-service stop with computerized pumps. They allow you to purchase only twenty shekels' worth— less than five dollars—of gas with any single credit card during a twenty-four-hour period. I purchase an initial dose with my Visa, but I am not sure it's going to be enough to get me to the next gas station. Gas is expensive here, so twenty shekels' worth is not much.

I'm just about to offer the pump my American Express card when a damp Palestinian man in a patched sky blue sweater approaches me. He has a hesitant, hopeful smile on his face and holds up a large scratched can with green Arabic lettering. It once held olive oil.

"My pickup truck ran out of gas on the road below here, and I don't have a credit card." He pulls a twenty-shekel note out of his pocket. "Do you think you could buy me some gas with your card?" I want to help, but my instincts tell me to be suspicious. Gas is not just for cars. It can be used to make Molotov cocktails and to torch houses. I can't see the road where his truck is supposedly stalled. Still, I very nearly ran out of gas on the same road myself. Do I really want to acknowledge that the conflict is so bad that Palestinian and Jewish motorists can no longer help each other out? Like most people, I prefer to live in the world as I want it to be rather than as it most probably is. I insert my card into the machine's slit, and gasoline gushes into the can. He thanks me, gives me the twenty-shekel note, picks up the can, and starts walking. I close my own gas tank, climb in my rented Hyundai, and head out of the gas station.

The man, whose name I didn't even ask, hobbles under the weight of the fuel. My best guess is that he's a few years younger than I, maybe forty, but he's a bit flabby and seems to have a limp.

Thirty meters after I pass him, just before I turn back onto the Alon road, I slam on my brakes, close my eyes, and rest my head on the steering wheel. I have little more than an hour to get back home before the Sabbath begins. I look north and south on the main road and see no sign of the Palestinian's abandoned pickup truck. I could do nothing more senseless than allow an enemy alien carrying a deadly weapon into my car, but I feel compelled to go back. Of course, his innocuous demeanor is the perfect camouflage for a decoy. He could lead me along deserted, unfamiliar turnoffs in search of his truck. He wouldn't need a gun—only a cigarette lighter—to lead me into an ambush.

I'm sure there is a scientific explanation for why I feel so torn, but it won't help me decide what to do.

The sun reflects off the tops of the cloud that obscure the rift below. If a continent splits unseen, it means nothing at all. Mountains move, horsts rise, grabens fall, water and salt wend their way underground and then surface. Even God can't see, because God's mind exists only when a human mind can seek it. Since the first humans found the valley in their northward wandering, a million and a half years ago, they have fashioned its rocks, observed and eaten its animals, drunk its water. They have gazed at its mountain walls and bathed in its river. They have made it a border, a crossing, a pasture, a hunting ground, the hinterland of a city. They have erected standing stones and carved idols, written scrolls, lodged at inns, raised families, and waged wars. They have sought to comprehend the body of the rift valley's laws—its movements, its trees, birds, and beasts, and the relics left behind by other, earlier residents and travelers. They have attempted to find paradise. They have spun out friendships and feuds, doctrines and interpretations. They have written poetry, told stories, and captured myths that have wafted in from other places like migrating birds. It is the stories that men and women have told to explain what they see and what they do as a result that create the rift as we see it. Since humans first began to call the names of gods, they have created their own valley of prayers, desires, deeds, and choices, which overlay the landscape just as the rain clouds do. As hard as we try to comprehend the landscape itself, it is humanity that we find.

I back up and meet the Palestinian halfway. "Get in," I say, forcing a smile. "I'll take you to your truck." It's an infinitesimal instant in the movement of the plates, a tiny act of enigmatic altruism on the edge of this crack in the earth.

NOTE ON SOURCES

One of the pleasures of writing this book was the opportunity to spend many hours in the libraries of the Hebrew University of Jerusalem and the Israel Antiquities Authority. A fair portion of that reading was devoted to learning about peoples, events, places, and controversies that in the end make no appearance in my story. For those interested in reading further about the subjects that do appear, I offer a very partial guide to some English-language sources I found useful. In some instances I cite the English versions of works I consulted in Hebrew.

I. FACTS ON THE GROUND

The song sung by Sarit Hadad from the boom box on the Eilat beach was written by Lior Farhi.

Uzi Avner presents his analysis of standing stones and their relation to ancient desert cults in his Hebrew University Ph.D. dissertation (2002), "Studies in the Material and Spiritual Culture of the Negev and Sinai Populations, During the 6th–3rd Millennia B.C."

I learned about Diocletian's reign from Stephen Williams, *Diocletian and the Roman Recovery* (Routledge, 1997). The inscription found at Kibbutz Yotvata is deciphered in Israel Roll, "A Latin Imperial Inscription from the Time of Diocletian Found at Yotvata," *Israel Exploration Journal* 39 (1989), pp. 252–60. The early results of the current excavation are summed up in Jodi Magness, Uzi Avner, and Gwyn Davies, "The Roman Fort at Yotvata, 2003," *Israel Exploration Journal* 54 (2004), pp. 256–61. The story of 'Abd al-Masih, the monk who was unmasked at Ghadyan, is translated and glossed in S. H. Griffith, "The Arabic Account of 'Abd al-Resch an-Nbagranial Ghassin," *Le Muséon: Revue d'étude orientale* 98, no. 34 (1995), Louvain-La-Neuve.

Amotz Zahavi's theory of evolution is presented in the very readable Amotz and Avishag Zahavi, *The Handicap Principle: A Missing Piece of Darwin's Puzzle* (Oxford University Press, 1997). Zahavi's handicap principle is not widely accepted in current work on the puzzle of altruism and reciprocity. For a summary of recent thinking and research, see Martin A. Nowak and Karl Sigmund, "Evolution of Indirect Reciprocity," *Nature*, 237, no. 27 (October 27, 2005), pp. 1291–98.

There are several good introductory works on the geology of the Dead Sea rift valley and the Levant, but all the ones I found are in Hebrew. Aharon Horowitz's magnum opus, *The Jordan Rift Valley* (A. A. Balkema, 2001), is largely technical, but the introduction and conclusion provide an accessible summary.

II. HANGING BY A HAIR

Jodi Magness's book on the Qumran excavations, *The Archaeology of Qumran and the Dead Sea Scrolls* (Eerdmans, 2002), is clear and sober and deserves a wide readership. Yizhar Hirschfeld (who, sadly, passed away while this book was being prepared for publication) presents a very different analysis of the same evidence in *Qumran in Context: Reassessing the Archaeological Evidence* (Hendrickson, 2004). Yosef Garfinkel, together with Michelle A. Miller, sums up the excavations at Sha'ar HaGolan in *Sha'ar HaGolan: Neolithic Art in Context* (Oxbow, 2002).

III. FLOATING IN THE AIR

I first read about the tragic love triangle of Berl Katznelson, Leah Meron, and Sarah Shmukler in Anita Shapira's *Berl: The Biography of a Socialist Zionist* (Cambridge University Press, 1984). The story of how Satan fooled Adam and Eve is translated and analyzed in Michael E. Stone, *Adam's Contract with Satan: The Legend of the Cheirograph of Adam* (Indiana University Press, 2001).

Elchanan Reiner presents his ideas about the Galilean traditions associated with Joshua and other biblical figures in "From Joshua to Jesus: The Transformation of a Biblical Story to a Local Myth, A Chapter in the Religious Life of the Galilean Jew," in Arieh Kofsky and Guy G. Strousma, eds., *Sharing the Sacred: Religious Contacts and Conflicts in the Holy Land* (Yad Itzhak Ben Zvi, 1998), pp. 223–71. To learn more about the scurrilous *Chronicle of Jesus*, I recommend Hillel I. Newman, "The Death of Jesus in the *Toledot Yeshu* Literature," *Journal of Theological Studies*, n.s., 50, pt. 1 (April 1999). John MacGregor's description of Tiberias comes from his account *The Rob Roy on the Jordan: A Canoe Cruise in Palestine, Egypt, and the Waters of Damascus* (J. Murray, 1904).

The battle of the Horns of Hittin is recounted in numerous sources; the details remain a subject of debate. A good summary is Benjamin Z. Kedar's "The Battle of Hattin [*sic*] Revisited," which can be found in a volume he edited, *The Horns of Hattīn* (Variorum, 1992). My quotations from Arab sources on the battle come from Francesco Gabrieli, *Arab Historians of the Crusades* (University of California Press, reprint, 1984), as does the story about the curiously calm cuckold of Nablus.

To see how Shmuel Marco correlates archaeological evidence with the historical record of earthquakes, see Shmuel Marco et al., "Archaeology, History, and Geology of the A.D. 749 Earthquake, Dead Sea Transform," at geophysics.tau.ac.il/personal/shmulik/GK-Geology2003.pdf, and Ronnie Ellenblum et al., "Crusader Castle Torn Apart by Earthquake at Dawn, 20 May 1202," *Geology* 26, no. 4 (April 1998), pp. 303–306.

Naama Goren-Inbar's papers on her finds at Gresher Benot Ya'aqov are technical, but an interested lay reader can benefit from looking at these primary sources. I recommend in particular the following (in all these she is the first named author, along with others): "A Butchered Elephant Skull and Associated Artifacts from the Acheulian Site of Gesher Benot Ya'aqov, Israel," *Paléorient* 20, no. 1 (1994), pp. 94–112; "Pleistocence Milestones on the Out-of-Africa Corridor at Gesher Benot Ya'aqov, Israel," *Science* 298 (August 11, 2000), pp. 844–947; "Evidence of Hominin Control of Fire at Gesher Benot Ya'aqov, Israel," *Science* 304 (April 30, 2004), pp. 725–27.

ACKNOWLEDGMENTS

When a journalist presents the work of researchers to the general public, he inevitably makes scholarship sound easy. By the time there's a headline for the journalist, the scientist or archaeologist or historian has organized the findings, marshaled the evidence, and produced a theory—a theory of being, for all intents and purposes, a story that has hard facts behind it. Journalists are trained to put together a story in a few days or weeks (or hours, even minutes, if need be). In contrast, a published article or book by scholars, such as Amotz Zahavi, Jodi Magness, Elhanan Reiner, and Naama Goren-Inbar, is the product of years of digging, examining, analyzing, and thinking.

So I am continually surprised anew, and eternally grateful, when scholars make the time and effort to explain their work to me. I am even more grateful in the case of this book, because the scholars knew that I would be using the results of their years of painstaking research for my own literary purposes. Obviously, the way I present their work here, the aspects that I stress and the contexts in which I place it, may be very different from the way they would present it themselves. Any errors are my responsibility alone.

In addition to the people mentioned by name in the text, a number of others provided useful pointers and suggestions: Donald Tzvi Ariel, Benny Begin, Haim Goren, Yossi Leshem, and Aren Maeir. If I have inadvertently forgotten others, my profuse apologies.

I encountered the work of Michael Stone and Naama Goren-Inbar on assignment for *The Chronicle of Higher Education*, where earlier versions of the material presented here appeared.

My father, Sanford Watzman; my sister, Nancy Watzman; and my friend Neal Feigenson read through drafts of the book and made many important suggestions. Kaddish Goldberg kindly read through and suggested corrections to the section on Tirat Tzvi. Nancy and Jeff Heller provided essential support. My agent, Simon Lipskar, enabled me to get the idea off the ground. Eric Chinski edited with rigor and sensitivity, and it was a pleasure to have Gena Hamshaw dealing with all the technical aspects of the publication process. The members of my book club read the galleys and caught a number of errors and inconsistencies.

My wife, Ilana, and my children, Mizmor, Asor, Niot, and Misgav, provide me with the love and support I need in order to write.

In a very fundamental way, the curiosity and respect for different ways of understanding that lie at the foundation of this book are the product of my education and upbringing, so with great love I have dedicated this book to my parents, June and Sanford Watzman.